John Shaw Billings, Edward Hammond Clarke

A Century of American Medicine from 1776 to 1876

John Shaw Billings, Edward Hammond Clarke

A Century of American Medicine from 1776 to 1876

ISBN/EAN: 9783337779085

Printed in Europe, USA, Canada, Australia, Japan

Cover: Foto ©berggeist007 / pixelio.de

More available books at **www.hansebooks.com**

A CENTURY

OF

AMERICAN MEDICINE.

1776—1876.

BY

EDWARD H. CLARKE, M.D.,
LATE PROFESSOR OF MATERIA MEDICA IN HARVARD UNIVERSITY, ETC.;

HENRY J. BIGELOW, M.D.,
PROFESSOR OF SURGERY IN HARVARD UNIVERSITY, ETC.;

SAMUEL D. GROSS, M.D., LL.D., D.C L. Oxon.,
PROFESSOR OF SURGERY IN THE JEFFERSON MEDICAL COLLEGE, PHILA., ETC.;

T. GAILLARD THOMAS, M.D.,
PROFESSOR OF OBSTETRICS, ETC., IN THE COLLEGE OF PHYSICIANS
AND SURGEONS, NEW YORK, ETC.;

AND

J. S. BILLINGS, M.D.,
LIBRARIAN TO THE NATIONAL MEDICAL LIBRARY,
WASHINGTON, D. C.

PHILADELPHIA:
HENRY C. LEA.
1876.

Entered according to the Act of Congress, in the year 1876, by

HENRY C. LEA,

in the Office of the Librarian of Congress. All rights reserved.

PHILADELPHIA:
COLLINS, PRINTER,
705 Jayne Street.

PUBLISHER'S NOTE.

The following papers, during their publication in the "AMERICAN JOURNAL OF THE MEDICAL SCIENCES," have met with commendation so general, that it has seemed due to the profession whose development for the last hundred years is here traced with so much fidelity, to place them in a form more convenient for future preservation and reference. Taken as a whole, they present a complete and connected review of the progress of medical science in America during the whole of the period in which medicine can be considered to be a science; and the volume, it is therefore hoped, will possess not only interest in the present, but an enduring value in the future.

PHILADELPHIA, October, 1876.

PRACTICAL MEDICINE.

BY

EDWARD H. CLARKE, M.D., A.A.S.,
LATE PROFESSOR OF MATERIA MEDICA IN HARVARD UNIVERSITY.

PRACTICAL MEDICINE.[1]

When Boerhaave, the most accomplished and celebrated physician of the 18th century, died, he left behind him an elegant volume, the title-page of which declared that it contained all the secrets of medicine. On opening the volume every page, except one, was blank. On that one was written, "keep the head cool, the feet warm, and the bowels open." This legacy of Boerhaave to suffering humanity typified, not inaptly or unjustly, the acquirements, not of medical science, but of medical art at the close of the 18th century. Empiricism, authority, and theory ruled the medical practice of the world at that time. The result of therapeutical experience from Hippocrates to Boerhaave was fairly summed up by the latter in the eleven words we have just quoted. To quiet the nervous system, to equalize the circulation, to provide for the normal action of the intestinal canal, and to leave all the rest to the *vis medicatrix naturæ* was sound medical treatment, and it was as far as a sound therapeutics had gone a hundred years ago. This goal had been reached by empiricism. Wise practitioners like Boerhaave, Sydenham, Morgagni, and a few others, were content to restrain their materia medica within these modest limits. The vast majority of practitioners, however, either blindly followed

[1] The author desires to acknowledge his indebtedness to Dr. R. H. Fitz, Assistant Professor of Pathological Anatomy in Harvard University, for invaluable aid in collecting many of the data upon which this essay is founded.

authority of the past, and bled and dosed by the book, or adopted some strange theory of planetary influence, signatures, animal spirits, or occult force, and treated disease in accordance with whatever theory they chanced to believe in. Medical practice, as a rule, deserved the ridicule of Molière and the satire of Montaigne

In making these statements we do not forget that there had been real progress in many departments of medical science. Anatomy, physiology, surgery, chemistry, and physics had made substantial conquests within their own domains. We do not forget that Harvey had discovered the circulation of the blood; that Haller, one of the greatest names in medicine, had discerned the fact of muscular irritability, and its connection with the nerves; that Albinus had introduced thoroughness and exactness, so far as the means and instruments of observation accessible to him rendered them possible, into anatomical investigation; that Morgagni had founded the science of pathological anatomy, which has since yielded such magnificent results; that Astruc in 1743 had announced the reflex phenomena of the nervous system, which Prochaska before the close of the century more fully developed; that Boerhaave, Sydenham, Mead, Hoffmann, and Stahl had rendered good service to practical medicine; that Franklin and others had brought electricity, magnetism, and galvanism into the domains of science, though their relations to medicine and physiology were not then recognized; and that chemistry had entered upon a career of investigation which it has since followed with extraordinary success. But all these discoveries were in the nature of isolated facts. They were more like islands, surrounded by an unknown ocean, than like parts of a continent, intimately connected with each other and forming portions of a grand and systematic whole.

In spite of these achievements, however, theory, empiricism, and authority ruled the medical world at the close of the 18th and beginning of the 19th century. Let us look at some illustrations of this statement.

Cullen, who flourished during the middle of that century, reasoning from *a priori* considerations, founded his pathology and nosology upon pure theory. He not only did this, but he recognized the fact that he did so and defended himself in doing it. He declared it to be the duty of a philosophical inquirer in medicine to control his observations by his theories, and not his theories by his observations. In like manner he maintained that the medical practitioner should be guided at the bedside, less by the indications of nature than by theoretical considerations. Such was the attitude with regard to the theory and practice of medicine of one of the most philosophical thinkers and learned physicians of that period. He was by far the ablest of the solidists. His views met with general acceptance in England, and excited a great influence upon the medical opinions and practice of this country, and especially of New England. Much of the practice of our fathers and many of their medical opinions may be traced directly to Cullen. He was too often obeyed as a superior. Fortunate was it that the common sense and independence of American physicians often led them to refuse obedience to his authority and to follow the guidance of rational empiricism.

Brown's theory of medicine, which appeared not long after that of Cullen, is another illustration of the speculative tendency of medicine at that time. Brown was a man of less breadth, learning, and power, but of a more practical turn, than Cullen. His practical tendencies led him to base his system chiefly on therapeutics. Its pathology was essentially that of Cullen, and its physiology a misconception of the Hallerian notion of irritability. Its essential error was that it rested not upon facts, but upon assumptions. Its motive was a desire to substitute a stimulating for a lowering method of treatment. Its practical characteristics, however, caused it to spread more rapidly and to exert a more profound influence over the medical opinions and practice of his time than that of his more philosophical contemporary. It rapidly made its way into Germany, France, and Italy. Dr. Rush, of Philadelphia, illustrious as a practitioner, writer, and signer of the Declaration of Independence, who was a disciple of Brown, imported it into America. Introduced under such auspices, it spread rapidly throughout the country and produced a deep and lasting impression upon American medicine.

The speculative tendency to which we have referred, found its most extravagant expression and attained its largest development in the theory which Hahnemann framed near the beginning of the present century. Although, on account of its manifest absurdities, it was rejected by all scientific men, yet, to the philosophic student of the history of medicine for the past hundred years, it is interesting, not only as a curious instance of the aberration of the human intellect, but because, without contributing at all to the progress of medical science, it has modified the therapeutics of the present age by reminding the physician of the limits of his art, and of the great part which nature plays in the cure of disease. Hahnemann ignored all previous medical knowledge. He denied that medicine was a branch of natural science; that any knowledge of anatomy, or physiology, or pathological anatomy, or of diagnosis, or of the investigation of the nature of disease was necessary to the physician; and also denied the existence of any curative power in the human system. Consciously, or unconsciously, abstracting from the mediæval doctrine of signatures its guiding principle, that the like colour cures the like colour, he declared that like cures like, *similia similibus curantur*. To this he added the doctrine of the potency of dilutions, and later admitted certain diseases which he called psora, sycosis, etc., as modifying elements. Symptoms and groups of symptoms were all that were worthy of the attention of the physician, and these were to be treated by potencies in accordance with his fundamental theory. It is unnecessary to allude to the modifications which this theory has undergone at the hands of his disciples. It is sufficient for our purpose to recognize it as a sort of zymotic element in the progress of medical art in this country and Europe, and one which, notwithstanding its activity for a considerable period, is now declining.

Such was the condition of medical science and art at the close of the eighteenth century. A few great minds isolated from each other, slaves to no theory, emancipated from authority and dissatisfied with the results of empiricism, busied themselves with the accumulation of facts whose value they scarcely recognized, but which the future would gladly use. Others, and a larger number, were framers, or disciples, or advocates of some sort of theory, whose foundations were almost purely hypothetical. The vast majority of practitioners, slaves of a routine which authority had sanctioned, were guided solely by empiricism. The outlook was by no means cheerful. It was evident that if medical science was to advance, and a rational therapeutics ever to become possible, some new element, or force, must be introduced. Fortunately this new element, or force appeared. It was introduced by two men, John Hunter of England, and Bichat of France, who may be justly called the founders of modern physiology and pathology.

John Hunter was one of those remarkable men who only appear at rare intervals, and who, if they enter the arena of politics, mould the fate of an empire; if that of theology, change the faith of the age; if that of science, enlarge the boundaries, and add to the sum of human knowledge. He recognized that medicine was one of the natural sciences, more or less intimately connected with all of them, and to be studied as they are by rigid and careful observation.

Theory was useless, except so far as it rested upon facts. He regarded a knowledge of the whole organic and inorganic world as necessary to a just comprehension of the structure and functions of man. "He determined to contemplate nature as a vast and united whole, exhibiting, indeed, at different times different appearances, but preserving, amidst every change, a principle of uniform and uninterrupted order, admitting of no deviation, undergoing no disturbance, and presenting no real irregularity, albeit to the common eye, irregularities abound on every side."[1] With such an object before him, he proceeded to collect data of every kind. The Hunterian Museum in London testifies to his indefatigable industry, and to the extent and accuracy of his researches. His method of investigation and of reasoning has served as a model for the age that followed him. His influence upon American medicine was not less potent than upon that of England and the rest of Europe.

At the time Hunter was at work in London, his great contemporary, Bichat, was engaged in those researches in France which have rendered such inestimable services to physiology. Bichat died young, but he lived long enough to show that he was one of the world's greatest minds. "Between Aristotle and Bichat," says Buckle, "I can find no middle man." He and Hunter represent the turning-point in medicine from idealism, speculation, and theory, to accurate and close observation. His great merit lay in recognizing the fact that power depends on structure, and the additional fact that a knowledge of structures can only be obtained by studying the formation of the tissues that compose them. By following the method of Bichat, Agassiz was led to the remarkable discovery of the intimate connection of the tegumentary membrane of fishes with their

---

[1] Buckle's History of Civilization in England, vol. ii. p. 446, Am. ed.

whole organization; by the same method, Cuvier, Owen and others ascertained the intimate relation of the teeth of an animal to its whole organization. The great discoveries in physiology of the past hundred years, are due to the fidelity with which physiologists have substantially followed the line of investigation marked out by Bichat and Hunter.

The American Revolution, which was the forerunner of political changes of the gravest character in Europe as well as in America, was coincident with this new departure in medicine. American medical science was necessarily an offshoot from that of Europe. While it inherited the traditions, the superstitions, the theories, the authority, and the empirical results of Europe, it also gratefully welcomed the independent thought and sound method of Hunter and Bichat. William Hunter's magnificent work on the gravid uterus (which for accuracy and completeness has never yet been surpassed) appeared in 1774. It was an admirable example of the results of careful investigation, and was a most auspicious illustration of what the new century was to accomplish. From that time to this the progress of medicine in all its branches has been of the most gratifying character. Although it is true, as Tennyson says, that

"Science moves but slowly, slowly creeping, creeping on from point to point,"[1]

yet, as we look back upon the past hundred years, we find that its march has been one of extraordinary rapidity. During this period, more of nature's great resources have been discovered, and more of her secrets found out than ever before. A thousand doubtful suggestions have ripened into facts. The telegraph, the locomotive, the steamship, the photograph, the spectroscope, and other discoveries more than we can enumerate, testify to the century's scientific

[1] Locksley Hall.

activity. In politics this century has witnessed "the separation of America from Great Britain, the formation of the United States, the meeting of the Tiers-États, the revolution, the downfall of the French monarchy, the republic, the rise of Napoleon, the mighty European wars which altered the face of Europe and ended with the 'Hundred Days' and the exile of the Corsican tyrant, the restoration of the Bourbons and their ruin, the Monarchy of July, the Second Republic, the Second Empire, the Third Republic, the Commune, and the humiliation of France by a power which but the day before had been a mere federation of incoherent atoms, the Septennate, the unification of Germany and of Italy."[1] In theology all faiths, from that of Catholic Rome to that of the latest Protestant sect, have been attacked, and they themselves have given unmistakable signs of hesitancy and change. The faith of Christendom has been, and is, crystallizing into new forms, and moving to new issues. It is not an extravagant assertion to say that in all this turmoil, change, and progress, medicine has kept abreast of the other natural sciences, of politics, and of theology, and has made equal conquests over authority, error, and tradition.

If this statement seems extravagant, it is to be recollected that the brilliant discoveries in natural sciences and the arts, the great political changes, and the vacillations of long-established faiths to which we have referred, influence so obviously the fate of nations and the aspects of civilization, that they force themselves prominently upon our attention, while the progress of medicine is silent and unobserved. Yet the progress and changes of the latter are not less real than those of the former, and, perhaps, affect more profoundly than they, the development of civilization and the welfare of the human race.

During the past century, medicine has been enfranchised

---

[1] The Nation, August 19, 1875.

from superstition, quasi-charlatanism, bald empiricism, and speculation, and has developed into a symmetrical science, affiliated with the other natural sciences, studied by the same methods and the same appliances as they are, and, like them, has been planted upon the solid basis of fact and demonstration ; pathological anatomy, starting from the de Sedibus of Morgagni and the labours of Baillie, and illustrated by the later researches of Rokitansky, Cruveilhier, Virchow, Recklinghausen, Cohnheim, and others, has become a fundamental branch of medical science ; obstetrics, rescued from the hands of ignorant midwives, has been raised with its allied branch, gynæcology, to its legitimate position as a science ; preventive medicine and hygiene, cultivated to an extent previously unknown, have prolonged the average of human life ; organic and physiological chemistry have been substantially created, and achieved important and brilliant results ; physiology, guided by Blumenbach, Magendie, Legallois, Dumas, Flourens, Johannes Müller, Carpenter, Schiff, Helmholtz, Claude Bernard, Hammond, Dalton, Flint, Weir Mitchell, and others, has grappled with the abstrusest problems of structure and life, and has revealed so much as to make timid people tremble at the audacity of its efforts ; the reflex action of the nervous system, first discovered by Astruc and Prochaska, has been shown by the admirable investigations of Sir Charles Bell, Magendie, Marshall Hall, Claude Bernard, Brown-Séquard, and their associates, to be, next to the discovery of the circulation of the blood, the most important addition to physiological knowledge that has yet been made—one that has illustrated and explained the complex and almost inexplicable nature of the nervous system; the inhibitory and vaso-motor system of nerves has, in part, been discovered ; the velocity with which sensation, thought, and volition are transmitted along the nerves has been measured and determined ; the automatic action of the nervous system, and the position of the ganglia as centres of nervous power, have been demonstrated; the secrets of digestion and assimilation have been disclosed; by a method of exploration, which Auenbrugger and Laennec discovered, and Louis improved, and Skoda has shown to be in harmony with the laws of acoustics, the interior of the chest has been laid open to examination, so that the condition of the lungs and heart can be marked out with an accuracy like that with which the engineer maps out the topography of a mountain; the natural history of some of the gravest diseases has been ascertained, and means of preventing or curtailing them discovered; the ophthalmoscope has revolutionized ophthalmology; the microscope has penetrated the secrets of structure and tissue; the spectroscope has traced the devious wandering of drugs from the stomach to the remotest organs of the body; the sphygmograph has revealed the unseen and delicate movements of the heart and pulse; the æsthesiometer has measured the sensitive power of tissue and nerve; the dynamometer has recorded the force of the muscles; chemical analysis has traced the transformation of food into various forms of force, such as motion, heat, and thought; the materia medica has been made rational and effective by cleansing it from the disgusting animal excreta and filthy compounds that defiled it, from the absurd farragos and useless formulas that superstition or theory had foisted into it, and by adding to it numerous agents that botany and chemistry have discovered; last of all, and most important of all, the grandest discovery of the ages, that which will render this century remarkable for all time, a class of anæsthetic agents has been discovered by which surgery and death even are deprived of half their terrors, and the physician at his will enabled to compel pain to disappear and distress to be quiet.

Such has been the progress, and such are some of the achievements of medical science for the past century. They are enough to justify the enthusiastic regard in which physicians hold their profession, and enough to deserve, as they have received, the gratitude of mankind. After this survey of the general progress of medicine for the past hundred years, we are prepared to estimate more correctly than would otherwise be possible, the part which the United States has taken in aid of this progress and in attaining these results.

In making up our estimate, however, let us remember that a large amount of scientific work cannot justly be expected of the medical profession in a new country. When the nation had acquired its independence, its population extended along a narrow coast-line from what was then known as Massachusetts, now Maine, to Georgia. The inhabitants had the Atlantic Ocean in front of them, and in their rear the unexplored forests, filled with aborigines, that stretched far away towards the Pacific. As a matter of necessity they were obliged to occupy themselves almost exclusively with the task of obtaining a secure existence in a new country. For the first fifty years of the nation's life, the necessities of the present left little leisure for the cultivation of the arts and sciences. The medical profession were compelled by their position to devote themselves almost, if not quite exclusively to the practice of their profession, and to leave scientific investigation and discovery to a later period. There was no superabundance of educated physicians. If Boerhaave, Cullen, Hunter, or Bichat had found themselves in America at that time, they would have been obliged to take care of the sick, rather than investigate the laws of disease and of life, and the world would not have heard of them as original investigators and natural philosophers.

Over fifty years ago, Sydney Smith, alluding to the slow progress of intellectual development in the first half of our national existence, said in the *Edinburgh Review* :—

"The Americans are a brave, industrious, and acute people, but they have hitherto made no approaches to the heroic, either in their morality or their character. During the thirty or forty years of their independence they have done absolutely nothing for the sciences, for the arts, for literature, or even for the statesmanlike studies of politics and political economy. . . . In the four quarters of the globe, who reads an American book? or goes to an American play? or looks upon an American picture or statue? What does the world yet owe to American physicians or surgeons? What new substances have their chemists discovered, or what old ones have they analyzed? What new constellations have been discovered by the telescopes of Americans? What have they done in mathematics? Who drinks out of American glasses, or eats out of American plates, or wears American coats or gowns, or sleeps in American blankets?"

It must be confessed there was a great deal of truth in his statements at that time. Naturally enough his words rubbed the backs of all loyal Americans the wrong way, and everybody cried out accordingly. At the present time we can read his biting language with equanimity. If the first half century of our national existence did not yield much to science and art, it produced all that could have been justly expected of it; and the last half has produced books, manufactures, discoveries in the arts and sciences of every kind that have gone over the four quarters of the globe. We can now fairly ask, Who does not read an American book? and can point with honest pride to the services which American physicians and surgeons have rendered to the world.

When Sir Humphry Davy was asked what he considered to be his greatest discovery, he replied, Faraday. In like manner we can justly say that American physicians and surgeons are the best contribution of the United States to medical science and art. The work which the physicians of the first age of the republic performed, and the way in which they performed it, proved them to be men of whom the nation need not be ashamed. Men like Rush, Physick, and

Chapman, of Philadelphia, Hosack, Watson, Francis, and Mott, of New York, the Jacksons, Warrens, and Bigelows, of Boston, Dudley, of Kentucky, and many others whom our space does not permit us to name, are contributions to science of the best sort. To the example and stimulus of their lives and work, may be justly ascribed, to a very considerable degree, the honourable position, acknowledged zeal, practical judgment, and sound attainments of the American medical profession of the present day. We have already referred to the intimate connection that existed a hundred years ago, and that fortunately still exists, between the medical science of Europe and of this country. The latter is not different from the former. The two are parts of a common whole. Even the war of the revolution scarcely disturbed this connection. An illustration of it is to be found in the fact that in the same year, 1796, in which Jenner vaccinated his first patient, Dr. Waterhouse repeated the operation in Cambridge, Massachusetts, and Dr. James Jackson in the neigbouring city, Boston. Another illustration of the same thing is shown in the education of American physicians. From the era of the revolution until now a large and constantly increasing number of American physicians, after having completed the curriculum of medical study in this country, have resorted to European schools for the completion of their professional preparation. Dr. Samuel Bellingham, who graduated at the first commencement of Harvard College in 1642, afterwards obtained the Doctor's degree at Leyden.[1] The best American education has always consisted in getting the best medical instruction that Europe and America jointly impart. Our medical schools are an honourable contribution to the medical work of the century.

We learn from Dr. Carson's History of the University of

---

[1] Historical Address. Dr. J. B. Beck, New York.

Pennsylvania, that the first course of medical lectures given in Philadelphia (and probably in this country) was delivered by Dr. Cadwalader, prior to 1751. The first systematic courses of lectures on medical subjects were given in Philadelphia a little more than one hundred years ago by Drs. Morgan and William Shippen, who were the fathers of medical teaching in America. The degree of Bachelor of Medicine was first conferred in Philadelphia in 1768, and that of Doctor of Medicine in New York in 1770. From these small beginnings sprang the medical colleges, which have ripened into the large institutions of Philadelphia, New York, and Boston, and into numerous other medical schools, too many we fear for the good of the profession and of the country, that are to be found in most of the cities, and connected with many of the colleges of the Union.

These medical schools were not founded by the State, nor are they controlled or supported by it. A few and only a few of them have been scantily endowed by private individuals. Their support depends upon the fees derived from the students that resort to them. They were called into existence by the necessities of the times when they were established, and from one decade to another have been modified in their organization and methods of instruction so as to meet the demands made upon them. They are the natural and necessary growth of circumstances. It would be an interesting and easy matter to trace them from the small beginnings that we have indicated to their present proportions, and to point out the law that has governed their development; but our limits permit only the briefest possible exposition of it.

During the colonial period, and for some time after the establishment of the republic, medical students derived their professional training, not from schools or universities, but from practitioners of greater or less eminence, with whom, to use a technical phrase, they entered their names as apprentices or students. By this arrangement they had the use of the library of their master, whose shelves, if not abundantly supplied, generally held a few books, and whose house usually contained in some closet or nook a few bones of the human frame, or perhaps an entire skeleton. These the student handled, examined, and studied. His opportunities for clinical study consisted in witnessing, and often assisting in the office practice of his master. There he pulled his first tooth, opened his first abscess, performed his first venesection, applied his first blister, administered his first emetic, and there first learned the various manipulations of minor surgery and medicine. After a time his clinical opportunities were enlarged by visiting with his teacher the patients of the latter, and becoming acquainted, not in hospitals but in private houses, with the protean phases of disease. His clinical lectures were his master's talk on the cases they had visited as they rode from house to house. After three years spent in this sort of study and practice, the young man was supposed to have acquired enough medical knowledge to enable him to commence the practice of his profession. In proportion as a physician or surgeon became eminent, students who had the means to do so flocked to him, and he became the centre of a medical school. His clinical instructions, instead of being the talks that beguiled the way of a long ride, were changed into formal lectures delivered in his study or in some private room. Those who proved to be the most popular teachers, and who lived in the same city or neighbourhood, associated themselves together for purposes of teaching. Thus were founded the medical schools of Philadelphia and other cities. They did not give, and were not intended to give, a complete medical education, but only to supplement the instruction of private teachers. The courses of lectures were few in number and brief in extent. Students still continued to enter their names, and study for the major part of the year with some medical man in their own neighbourhood, and to attend lectures, as it was called, only three or four months of the year. Gradually a larger demand was made upon the schools; their lecture terms were lengthened; professorships were subdivided; new ones were added; hospitals were utilized for clinical instruction; the schools continued to enlarge their curricula of study, and at length added summer instruction to their winter's work; museums were established; chemical laboratories were formed; microscopical departments created; and all the appliances were attached to schools that are necessary in the investigation of structure, life, and disease. This process of growth has not yet stopped. It is still going vigorously on. One university, Harvard, requires all its medical students to go through a systematic course of training, under its own supervision, by a corps of teachers of its own appointment.

It is evident, from this brief sketch of the medical schools of the United States, that they are different in their organization, and to a considerable extent in their objects, from those of Europe. It is equally evident that the former are gradually approximating the latter, though it is not likely that their organization, methods of instruction, and character will ever be the same. The fact that the European schools are founded and controlled by the State, and are to a large extent responsible to it, and that American schools are independent institutions, self-supporting, and responsible only to public opinion, necessarily impresses a distinctive character upon the medical schools of the two continents. The atmosphere of each is different; each leads a different life; and each will produce a different result. Admitting such to be the case, it does not follow that the medical schools of the United States are necessarily of an inferior character, or that the physicians who graduate from them are imperfectly educated. For the schools, except in the case of Harvard, just referred to, do not pretend to give a complete education, but only to supplement that which the student gets elsewhere. Indeed it may be affirmed that those who, like the apothecary of England and the Secundär Arzt of Germany, are charged with the medical care of the mass of the community in Europe, are not better equipped for the practical work of their profession than their average American contemporary. We do not mean to assert by this that the scientific training of our schools is equal to that of Vienna, Berlin, or Paris. But we do assert that if the necessities and different conditions of Europe and America are impartially compared, we shall find that the American method of medical education yields as good a practical result to the nation as the European method of medical education does to Europe. And we further assert that the flexibility of the American method permits of change, growth, and development, in correspondence with the demands of each succeeding age, more easily and more rapidly than is possible with the conservative organizations of Europe. Hence we are not ashamed to present our medical schools, with all their short-comings and imperfections, as substantial contributions to the practical medicine of the century. And, moreover, we can justly point to graduates of these schools, some of whom have, and others of whom have not, been fortunate enough to add to their American a European education, as in every way the peers of European physicians or surgeons.

It was a noteworthy and fortunate circumstance, that at the time of the establishment of the republic, the medical profession of the new nation contained a large number of intelligent, able, and well-educated physicians. Pre-eminent among these was Dr. Benjamin Rush, of Philadelphia, who devoted himself with enthusiasm to his profession, which he studied first in Philadelphia, and afterwards in Edinburgh. An ardent patriot, a lover of liberty, a friend of Washington, a signer of the Declaration of Independence, he was not only eminent as a physician, but distinguished as a philosopher and a scholar. Holding a high social position in a community, noted alike for its love of the arts and sciences, and for the graces of social life, he contributed largely to raise the profession of medicine in the estimation of the community in which he lived, and of the whole country. During the Revolutionary war he rendered essential service to the army by a variety of professional labours, and after its close remained permanently in Philadelphia. Notwithstanding the demands of a large practice, he found or made time for the investigation of scientific questions, and for the publication of the results of his inquiries. His treatise on Diseases of the Mind, regarded as a work full of instruction, and of great originality by Prof. Brown, of Edinburgh, contains many practical and original observations, and was a valuable contribution to psychological medicine. It is not yet forgotten. Dr. Tuke, in his late monograph upon the Influence of the Mind upon the Body, quotes from it approvingly. Speaking of another of the essays of Dr. Rush, Dr. Tuke says: "Rush wrote an able essay (and when are his essays not able?) on Hydrophobia, in which he assigns an important *rôle* to the influence of fear, and an involuntary association of ideas."[1] Few are the observers and writers whose labours are remembered and words quoted for a hundred years after they have ceased from their work. The observations of Dr. Rush on Yellow Fever were extensive and important. They produced an impression on both sides of the Atlantic. Although their pathology was erroneous and their therapeutics atrocious, they were a substantial contribution to medical science by the stimulus which they gave to the careful and exact study of disease. When Rush began his lectures as Professor of the Institutes and Practice

---

[1] Illustrations of the Influence of the Mind upon the Body in Health and Disease, by Daniel Hack Tuke, M.D., Am. ed., p. 202.

of Medicine in the University of Pennsylvania, diseases were divided, according to the nosology of Cullen, into orders, classes, genera, and species, containing about thirteen hundred and eighty-seven diseases, for each of which there was supposed to be an appropriate treatment. Rush rejected these arbitrary divisions. He paid little regard to the name of a disease, and founded his treatment on its nature and on the condition of the system. By this course he reduced his materia medica to a few active medicines, and so prepared the way for the simplification of remedies that has been accomplished since his day.[1]

Dr. Philip Syng Physick, a friend of Dr. Rush, and a favourite pupil of that great master, John Hunter, was one of the most accomplished and brilliant of American surgeons. He was not a prolific writer, but he found time, however, to study the character of yellow fever, and to publish the result of his observations, which were founded on post-mortem examinations. His researches into the character of this disease, together with those of Rush, La Roche, Alonzo Clark, Jones, and others too numerous to mention, form a library of yellow fever literature which will be more fully noticed in a subsequent essay, and which later investigators into its nature cannot afford to neglect.

While Dr. Rush was pursuing his investigations in Philadelphia, two men in Boston were labouring with equal zeal and earnestness in the cause of medical science. One of them, Dr. John C. Warren, devoted himself chiefly to surgery, and his work in that direction will be noticed in the surgical part of these memoirs. Apart from surgery he rendered a service to practical medicine that should not be forgotten. By his paper upon diseases of the heart, he first brought distinctly to the notice of the profession in this country that class of affections which Corvisart described in

---

[1] *Vide* Thatcher's Medical Biography.

his remarkable treatise. Another and more important service was the foundation and endowment of the anatomical museum of the medical department of Harvard University. Under his care and that of Dr. J. B. S. Jackson, who has worked in it and for it for more than a quarter of a century with rare intelligence and devotion, it has attained a completeness and excellence that few similar collections possess, and which render it one of the best contributions to the study and illustration of practical medicine in the country. In like manner, the large museums containing anatomical and pathological specimens, that have been collected in Philadelphia and New York and other medical centres of the United States, are invaluable contributions to the same science.

Dr. James Jackson, the second labourer to whom we referred, was known exclusively as a physician. He was one of the founders of the Massachusetts General Hospital, and, like Dr. Warren, was connected with the medical school of Harvard College at its commencement. He was a large practitioner, an acute and close observer of nature, but not a prolific writer. In him that indefinable but substantial something, called common sense, was applied with singular success to the practice of his profession, to his clinical teachings at the Massachusetts General Hospital, and to his didactic lectures at the medical school. His report on typhoid fever, and Dr. Hale's paper on the same disease, which may be found in the Communications of the Massachusetts Medical Society, were based on their own observations. The results at which they arrived were substantially those of Louis.

Dr. Jackson's Letters to a Young Physician are models of sensible advice to a practitioner whether young or old, and whether living on one side of the Atlantic or the other. He never indulged in heroic practice, or in therapeutic expedients for which he could not give a reason. He believed in the conservation of nature's forces. To a large extent the medical profession of New England was moulded by his teachings and example. The impression which he made is not yet effaced. Such an influence, though difficult to describe or estimate justly, is nevertheless a real contribution to practical medicine. Dr. Nathan Smith, a contemporary of Rush, Warren, and Jackson, deserves also to be remembered. He was a sound observer, who, having enfranchised himself from the bonds of authority, delighted to study nature with his own eyes, and was not afraid to follow where she led. His essay on Typhus Fever, published in 1824, had the merit of pointing out the self-limited nature of that disease, and of showing from his own experience the futility of attempting to abort it, or to treat it with violent remedies. "I have never been satisfied," he says, "that I have cut short a single case of typhus that I knew to be such. Typhus has a natural termination like other diseases which arise from specific causes." He mentions with approbation the successful treatment of a physician who gave only milk and water to his patients in this complaint. "All that is required," is Dr. Smith's therapeutical conclusion, "are simple diluent drinks, a very small quantity of farinaceous food, and avoidance of all causes of irritation." This result, which he reached by his own observations more than fifty years ago, is the same as that which has lately been loudly proclaimed in England and Germany. What Dr. Smith calls typhus was undoubtedly typhoid fever. At the time he wrote, typhus and typhoid fever were confounded together as different forms of the same disease. It is worthy of remark that Dr. Smith recognized the fact, now acknowledged, that typhoid fever arises from a specific cause, and that one attack of it prevents a subsequent one.

Typhoid fever prevails to such an extent in the United States, that our physicians enjoy ample opportunities for the study of it. Among those who have investigated it, none have done so with greater acuteness and ability than Dr. Gerhard, of Philadelphia, or have discriminated with greater clearness than he the essential differences between typhus and typhoid. He was the first, or among the first, to point out these differences with scientific accuracy. He says himself:—

"The advantages which I enjoyed of carefully studying the pathological anatomy and the symptoms of the two fevers, enabled me to place the question of their identity (typhus and typhoid), upon more settled scientific points, than had yet been done. . . . . . It is true that after the observations, which formed the basis of the paper which I published in 1837, were collected, but before their publication, Dr. Lombard, of Geneva, who was of course familiar with typhoid fever, stated in the *Dublin Journal* that the two diseases were different; the same remark I remember to have heard Prof. Andral make on the authority of Dr. Alison, and it was obvious to many persons that the description of Dr. Louis did not apply to the British typhus, but the points of resemblance and of difference were not settled, that is, they were not scientifically demonstrated."[1]

The merit of having decided this important question, of having demonstrated the essential difference between typhus and typhoid fever, belongs chiefly, if not wholly, to Dr. Gerhard, and so far redounds to the honour of American Medicine. Previous to his paper, which was published in the *American Journal of the Medical Sciences*, the evidence as to the essential distinction between the two fevers was mainly speculative, or conjectural;[2] he made it logical, clear, and unequivocal. It is only just in this connection to refer to the papers of Dr. J. Baxter Upham, of Boston, which, published many years after the appearance of Dr. Gerhard's

---

[1] A System of Clinical Medicine. By R. J. Graves and W. W. Gerhard, 1848, p. 735.

[2] *Vide* Am. Journal of Med. Sciences, vol. xix. p. 289, Feb. 1837; also, Wood's Theory and Practice of Med., vol. i. p. 373.

memoir, and founded on the careful personal investigation of Dr. Upham, confirmed the results of Dr. Gerhard, and added to our knowledge of the history of typhus. The observations of Dr. Thomas Stewardson on remitting fever form a valuable addition to our knowledge of that disease. The paper[1] which embodies his views was founded on the clinical study and post-mortem appearances of the cases which came under his notice in the Pennsylvania Hospital. In this memoir he calls attention to changes in the liver, which were present in every case, and were of a character not met with in other diseases. These he regarded as the anatomical characteristic, though not the primary seat of the disease.

Yellow fever has several times within the past century ravaged the Atlantic and the Gulf coast, so that our physicians have had unfortunately ample opportunities of studying the disease. Without detracting from the valuable labours of many other observers, it may be stated that to Dr. Deveze, then resident at Philadelphia, we are indebted for being foremost in asserting and maintaining the non-contagiousness of yellow fever; and to Dr. Alonzo Clark, of New York, for showing that the pathological change, so constantly observed, in the liver, is due to acute fatty degeneration.

Dr. S. H. Dickson, of South Carolina, had the opportunity of observing an epidemic of dengue, more than twenty-five years ago, of which he gave a highly interesting account. He considered the disease to be the same as that which prevailed at the South in 1828, and as the breakbone fever, described by Rush in 1778. The memoir is an instructive and valuable one.[2]

Dr. Gerhard's labours in practical medicine have contributed materially to its progress, and have given him a

[1] Am. Journ. Med. Sci., 1841 and 1842.
[2] Charleston Med. Journal, 1850.

deservedly high position among American medical scientists. Though our limits forbid an enumeration of all of his contributions, we cannot refrain from calling attention to his observations upon tubercular meningitis. Together with M. Rufz, he was the first to point out clearly the essential connection of hydrocephalus with tubercles of the pia mater, and the dependence of the former upon the latter.[1] Previous to his investigations, the notions of medical men with regard to the presence and cause of water within the cranium, were confused, theoretical, and consequently inaccurate. By many acute hydrocephalus was regarded as a cause, not as an effect—as an independent disease, not as a result. Dr. Gerhard cleared away the obscurity, supplied the missing links, and showed that tubercular disease of the meninges of the brain is a distinct malady which leads to the effusion of liquid there, as certainly as tubercle of the lung leads to purulent expectoration.

From the time of Hippocrates until recently the treatment of effusion into the pleural cavity has been among the *opprobria medicorum*. With the hope of promoting the absorption of the fluid, the unfortunate subjects of it were sometimes bled, *coup sur coup*, sometimes salivated with heroic persistence, often blistered with indefatigable zeal, generally plied with diuretics, and by cautious practitioners treated on the expectant method, and all with the result of not interrupting the progress of the effusion. In many, perhaps in the majority of cases, the powers of nature were equal to the demand made upon her and the liquid absorbed. In a large number of cases, however, this fortunate result did not occur, and the effusion went on increasing until the patient was killed by mechanical pressure, or by the development of some disease, like tubercle or other trouble that the

---

[1] Am. Journ. Med. Sci., xiii. p. 313; Wood's Practice vol. ii. p. 675.

pressure induced. More than a quarter of a century ago Dr. Henry I. Bowditch, of Boston, whose life has been devoted to the study of diseases of the chest, was impressed with the notion that it would be possible and safe to relieve this class of cases by drawing the fluid off. He made several attempts to do this by means of incisions into the pleural cavity. The results were not satisfactory. While Dr. Bowditch was busy with these efforts, Dr. Morrill Wyman, of Cambridge, who, unaware of Dr. Bowditch's views, entertained similar notions, successfully tapped a patient, by means of an exploring trocar and canula with suction-pump attached. In 1850, Dr. Bowditch, aided by Dr. Wyman, repeated the operation with equal success upon another patient, using the same apparatus. "That apparatus," says Dr. Bowditch, "I have modified somewhat, so as to make it, I think, more convenient; but the principle of the instrument remains as suggested by Dr. Wyman.'"[1]

From that time to the present, Dr. Bowditch has used his modification of Dr. Wyman's instrument for this operation. In his opinion it operates more rapidly than Dieulafoy's aspirator, and quite as harmlessly and easily for the patient. He has operated upon patients of all ages and both sexes, and with almost every species of complication, and has never seen any permanent evil results. His own statement is, that he has very rarely seen anything following the operation, but ease to the patient. During the last twenty-five years he has operated 325 times upon 204 persons. In a large number of these cases relief was not only afforded to the sufferer, but imminent death was prevented. This result is a demonstration not only of the propriety but of the necessity of performing thoracentesis in

---

[1] Thoracentesis, a paper read before the New York Academy of Medicine, April, 1870, p. 6, by H. I. Bowditch, M.D.

appropriate cases. Dr. Bowditch considers the following to be the indications for the operation:—

"1st. To save life when immediately threatened.
"2d. To prolong life, even when complicated with severe disease.
"3d. To shorten latent pleurisy.
"4th. To give temporary relief merely in absolutely hopeless cases.
"5th. To relieve cases of common pleurisy which do not easily yield to remedies after a few weeks of treatment."[1]

Thoracentesis is now regarded both in Europe and America as a legitimate, safe, and necessary procedure, when withdrawal of fluid from the chest is indicated. It has not won this position, however, without difficulty. It has had to run the gauntlet of opposition and of severe criticism from physicians and surgeons of great experience and reputation on both sides of the Atlantic. Trousseau advocated it; Valleix condemned it. English and American surgeons denounced it as unsafe and needless. That it has gradually made its way to its present acknowledged position, is largely due not only to the brilliant results of Dr. Bowditch's personal experience, but to the earnestness with which he has pressed by his pen the importance of it upon the profession, and the clearness with which he has pointed out the proper method of performing it.[2]

The principle of M. Dieulafoy's aspirator, an instrument too well known to need description, and lately introduced

---

[1] Thoracentesis, ut supra, p. 6.

[2] Those who are desirous of consulting Dr. Bowditch's papers on the subject, are referred to the American Journal of Medical Sciences, April, 1852, and Jan. 1863; American Medical Monthly, Jan. 1853, New York; Boston Medical and Surgical Journal, May 25, 1857. Thoracentesis and its General Results, address before the New York Academy of Medicine, April, 1870.

to the notice of the profession, is the same as that of Bowditch's exploring trocar and canula with suction-pump attached. The French physician's application of "aspiration" to all parts of the human body, is a brilliant generalization of the American physician's operation of thoracentesis. It is much to be regretted that M. Dieulafoy, in his admirable monograph on aspiration, neglected to make the slightest allusion to Dr. Bowditch's previous and persistent labours. Such a neglect on the part of M. Dieulafoy must have arisen either from an ignorance of Dr. Bowditch's previous investigations, or from a desire to claim and wear the laurels that another had won.

Consumption is recognized as the most terrible scourge of temperate climates. We are so familiar with its presence that we have ceased to be alarmed at its existence among us, although it causes from an eighth to a fifth of the total number of deaths in New England, and a very large proportion of all the deaths throughout the United States and Europe. The ablest intellects of the profession have occupied themselves, and are still occupied with the study of this disease, hoping to unravel completely its natural history and pathology, and to learn how to check its ravages and ameliorate the suffering it produces. Among these labourers Dr Bowditch holds an honoured place. His investigations led him to the conclusion that soil moisture is a large factor in the production and development of consumption. In May, 1862, he delivered an address before the Massachusetts Medical Society upon this subject.[1] In this address he was the first to announce what is now generally received as an acknowledged fact—that consumption may be produced in a family by residence on a damp soil. His language in the address referred to is as follows:—

[1] Medical Communications of the Massachusetts Medical Society, vol. x. No. 2, 1862.

"First. A residence on or near a damp soil, whether that dampness be inherent in the soil itself, or caused by percolation from adjacent ponds, rivers, meadows, marshes, or springy soils, is one of the primal causes of consumption in Massachusetts, probably in New England, and possibly in other portions of the globe.

"Second. Consumption can be checked in its career, and possibly, nay probably, prevented in some instances, by attention to this law."

The estimation in which these conclusions with regard to the influence of soil-moisture as a cause of phthisis, and of Dr. Bowditch's part in the investigation of it, may be inferred from the following statement: In 1867, Mr. Simon, of England, medical officer of the Privy Council, presented the results of Dr. Buchanan's investigation into the death-rate of towns in which soil-drainage had been introduced. The latter had ascertained that moist towns, in which this had been done, had a less death-rate from consumption after doing it than before. In consequence of this result, Dr. Buchanan was ordered to investigate thoroughly the subject. He made "an elaborate examination of the distribution of phthisis as compared with variations of the soil in the three southeastern counties of England." Mr. Simon concludes from this investigation, confirmed, as he states, by Dr. Bowditch's previous researches in America, "*that dampness of the soil is an important cause of phthisis to the population living upon that soil*" (italics as in the original). Mr. Simon adds, "this conclusion must henceforth stand among those scientific certainties on which the practice of preventive medicine has to rest."[1]

While these pages were passing through the press, a work on phthisis[2] appeared from the pen of Dr. Austin Flint, of

---

[1] Tenth Report of the Medical Officer of the Privy Council, 1868, p. 16.

[2] Phthisis; its Morbid Anatomy, Etiology, etc. etc. By Austin Flint, M.D. Phila., 1875, p. 441.

New York, which will be gladly welcomed by the profession of America and of Europe. It is based on a careful record of six hundred and seventy cases of phthisis, which are grouped and analyzed with reference to the practical deductions that may be legitimately drawn from them. The book is written from a clinical stand-point. So far as practicable, Dr. Flint follows the numerical method of investigation. For the most part the cases, which he reports, are chronic in their character, and belong to a class remarkable for the uniform character of the lesions, and of the symptomatic events and laws which are developed by their clinical history.

In addition to this recent work on phthisis, and to other labours, which we have elsewhere referred to, practical medicine is indebted to Dr. Flint for a great deal of valuable work. His reports on continued fever, and articles in the *American Journal of the Medical Sciences,* on Tuberculosis, Heart Sounds, Pneumonia, Chronic Pleurisy, have all of them deserved and received the careful consideration of the profession.

The progress of medicine, like that of all science, depends first upon the collection of facts, and afterwards upon a correct interpretation of them. Whoever recognizes a fact, however insignificant it may seem to him, and reports the discovery, makes a valuable contribution to science. The chief difficulty in the way of collecting accurate data, especially in medicine, is that few observers are gifted with the power of knowing a fact when they see it. "The hardest thing in the world, sir, is to get possession of a fact," said Dr. Johnson. Most observers report what they think to be, not what is. Whoever contrives a new instrument that increases the accuracy of physical exploration, whoever discovers a new method of examination, or modifies an old one, by which some secret of the organization is disclosed, whoever demonstrates the correct explanation of any phenomenon of the human system, whether it be the crackling of bubbles in the chest or the mechanism of thought in the brain, whoever traces back any symptom to its cause, so as to make the former the pathognomonic sign of the latter, or whoever in any way, by microscope, analysis, scalpel, or experiment, reveals anything that pertains to the structure or functions of man, in health or disease, contributes to the progress of practical medicine. It would be pleasant, if it were possible, to collect all the contributions of this sort, small as well as large, that have been made by Americans during the past hundred years to medical science and art. While the parentage of many of these contributions is well known and recognized, there are many others now incorporated into the body of science that cannot be traced to their discoverers; their lineage is unknown. The following pages record some of these contributions, in addition to what we have already described. We are sorry that we cannot make the record more complete than it is.

Dr. James Jackson, Jr., of Boston, whose premature death was not only a great personal bereavement to his friends, but a great loss to the science whose devoted student and servant he was, while pursuing his studies in Paris communicated in 1833 a paper to the Société Médicale d'Observation on the subject of a prolonged expiratory sound as an early and prominent feature of bronchial respiration, and one which frequently constitutes an important physical sign of the first stage of phthisis.[1] The accuracy of this observation has been demonstrated by many other observers since the appearance of his paper. At the present time a prolonged expiration, when heard in the clavicular region of the chest, is acknowledged as one of the earliest

[1] A Practical Treatise on the Physical Exploration of the Chest, etc. By Austin Flint, M.D., second edition. Philadelphia, page 191.

and most valuable signs to warn the practitioner of the insidious approach of disease. Probably few have ever heard even of the name of the young physician whose quick ear first caught the sound, and whose careful observation connected it with the condition that produced it.

When Laennec made his great discovery, which has revolutionized the study and indirectly the therapeutics of affections of the chest, a variety of stethoscopes were devised to conduct the sounds of that region to the ear of the observer. Most, if not all, of these instruments were clumsy and poorly adapted to the object in view. They gradually fell into disuse. Direct auscultation, by laying the ear directly on the chest, or with a single intervening bit of cloth, yielded a better result than the stiff, awkward wooden tube which Laennec employed, and which Dr. Holmes has so cleverly satirized. Dr. C. W. Pennock, of Philadelphia, while making his well-known investigations with regard to the heart and its diseases, discarded the stiff wooden instrument and introduced a flexible tube stethoscope.[1] Its advantages were obvious. It did not transmit the impulse, but only the sounds of the heart and chest, to the ear of the examiner. While using this instrument the physician was able to explore the sounds of the heart and chest undisturbed by any muscular movement. Dr. Cammann, of New York, improved upon Pennock's flexible stethoscope by adopting with some modifications the double binaural stethoscope of Dr. Arthur Leared, of London. This instrument conducts the sounds of the chest to the ear of the auscultator more clearly than any other, and does not conduct the impulse. It is the most serviceable stethoscope that has yet been devised.[2]

[1] Wood's Theory and Practice of Medicine, vol. i. p. 209.

[2] Dr. Arthur Leared, of London, exhibited at the great Exhibition in 1851 a double binaural stethoscope which he was the first

Dr. Alfred Stillé,[1] of Philadelphia, was among the first, if not the first, to call attention in print to a condition of the heart observed among soldiers as the result of prolonged and violent exertion, and now known as irritable heart; and Dr. Henry Hartshorne,[2] in the same year, more fully described the affection in a paper which he read before the College of Physicians of Philadelphia.

In a communication forwarded in December, 1862, to the Surgeon-General's Office, Dr. J. M. Da Costa[3] called attention to this same cardiac malady to which he gave the name of irritable heart, and his Medical Diagnosis, published in April, 1863, contains an outline sketch of the disorder. A few years later[4] he traced the connection of irritable heart with organic disease, and illustrated it with cases; in this paper, also, the inquiry took a wider scope and showed how exertion and strain could result in endocarditis and subsequent valvular disease, and in hypertrophy. In 1871 he published a careful and elaborate clinical study of irritable heart[5] based on upwards of 300 cases, in which he showed that irritable heart resulted from exhausting diseases, such as fevers and diarrhœa, and from strains and blows, as well as from muscular exhaustion, and further traced the connection between functional heart disorder and organic change. In it was also made a valuable contribution to a more exact

---

to devise. Dr. Cammann evidently got the idea of his instrument from that of the London physician, from which it differs in a few particulars.

[1] Address before the Philadelphia County Medical Society. Delivered Feb. 11, 1863, by Alfred Stillé, M.D.

[2] Am. Journ. Med. Sciences, July, 1864.

[3] Ibid., January, 1871.

[4] Sanitary Commission Memoirs, Medical Volume, New York, 1867.

[5] Am. Journ. Med. Sciences, January, 1872.

knowledge of the action of remedies on the heart.[1] In 1874 he[2] called attention to the same affection occurring with the same sequelæ in civil practice. In this brief monograph the effect of cardiac strain upon the muscular walls, valvular apparatus, and great vessels of the heart, is clearly stated, as well as the general symptoms and local signs. These papers give an excellent account of the disease they describe, and make a valuable and original contribution to practical medicine.

In a recent number of the *American Journal of the Medical Sciences*,[3] Dr. Da Costa has called attention to the advantage of forced respiration on the part of the patient as an aid to the physician in diagnosticating diseases of the chest. We can ourselves bear testimony to the accuracy of his statement. Forced respiration is of especial service in doubtful cases, particularly when it is important, as it often is, to make out a differential diagnosis between bronchitis and phthisis. It renders other services than this, for an account of which the reader is referred to the original article.

Dr. Da Costa has prepared, during the past twenty years, a number of papers, based upon his own observations of disease, which are valuable contributions to practical medicine. We regret that we are unable to do more than allude to some of them. In addition to those which are mentioned elsewhere, he published, in 1855, a memoir[4] on the pathological anatomy of pneumonia. In 1859 he published the results of some observations[5] "On the occurrence of a blowing sound in the pulmonary artery, associated with affections of the lung; on the sounds of the artery in health, and the effect on them and on the heart of the act of respiration."

[1] These papers of Dr. Da Costa, which were based upon his army experience, have lately received a German translation.
[2] On Strain and Over-action of the Heart. Toner Lecture, No. 3. Washington, 1874.
[3] July, 1875.
[4] Am. Journ. Med. Sciences, Oct. 1855.
[5] Ibid., Oct. 1859.

In 1866 he published a paper[1] on typhus fever, based upon the cases under his charge, and of course written from a clinical stand-point. In 1869 he gave to the profession a memoir on Functional Disorders of the Heart;[2] in which he attempted to show the real value and meaning of a cardiac murmur. In 1871 he recorded his observations on Membranous Enteritis,[3] which, like most of his other observations, were based on a careful clinical study of the disease, and present a complete account of it.

The importance of distinguishing the variations of pitch elicited by percussion is now universally recognized as an aid in ascertaining the condition of the organs in the chest. There are cases in which these variations afford the earliest clew to commencing disease; and sometimes when the signs are nearly evenly balanced it throws the vote which decides the verdict. The profession are indebted to Dr. Austin Flint, of New York, for calling their attention to this subject, at least in this country. Dr. Flint's statement of the value of variations of pitch in exploration of the chest, and the practical inferences from them which his acute observation and large experience suggested, and the investigations which led him to his conclusions in this matter, may be found in an essay which received in 1852 the prize of the American Medical Association. The combination of percussion and auscultation, or auscultatory percussion, as described and employed by Dr. Alonzo Clark, of New York, is undoubtedly well adapted to determine with ease and accuracy the boundaries of the heart.[4] This sort of cardiac examination cannot be made accurately without the aid of Cammann's stethoscope. When we recollect the method by which Piorry used to map out the boundaries of the heart, a task which

---

[1] Am. Journ. Med. Sciences, Jan. 1866.
[2] Ibid., July, 1869.   [3] Ibid., Oct. 1871.
[4] New York Medical Journal, July, 1840. Flint, on Diseases of the Heart, Second edition, p. 43, 1870.

we have often seen the distinguished French auscultator undertake in the wards of his own hospital five-and-twenty years ago, we are forcibly struck with the advance which has been made during the past quarter of a century in the physical examination of the chest.

There are few practitioners who have not sometimes been puzzled to distinguish between the solidification of pneumonia and the effusion of pleurisy. The differential diagnosis between these two conditions is sometimes a matter of great delicacy and difficulty. Here we are again indebted to Dr. Flint, of New York, for enabling us to solve the difficulty with comparative ease. He showed that by mapping out the lobar dulness which exists in pneumonia, the inflammatory condition of the lung could be discriminated from the effusion in which no such limited dulness exists. Though it does not fall within the scope of this paper to touch at all upon the subject of American medical literature, we cannot refrain from referring in this connection to the masterly digests of the vast number of memoirs, monographs, and the like, upon the subject of pneumonia, and perhaps we should add yellow fever, which have appeared from the pens of Dr. La Roche and Dr. Flint. They are substantial contributions to practical medicine. The mechanism by which the crepitant râle of pneumonia is produced is not yet perfectly made out. The explanation of it, given by Dr. E. Carr, of Canandaigua, N. Y., has been accepted by pathologists as probable, if not fully demonstrated, and deserves mention. Dr. Carr suggests that the crepitant sound is produced by air rushing into and distending the bronchial vesicles which had been previously glued together by tenacious mucus. For a full exposition of his views our readers are referred to his original article.[1]

[1] American Journal of Medical Sciences, New Series, vol. iv. p. 360, 1842.

Croup, a name dreaded alike by physicians and mothers, was for centuries the generic term of several inflammatory affections of the throat that were confounded together. Gradually these different affections have been discriminated from each other and have received different names. The term croup, or as some prefer to call it, membranous croup— the diphtherite of Bretonneau—is now restricted to an inflammation of the upper part of the air passages attended with the formation of a membrane. The membrane is recognized as an essential part of the disease. Richard Bayley, Surgeon of New York, recognized the distinctive characteristics of this affection as long ago as 1781. In a letter to William Thornton, M.D., of London, which afterwards appeared in the *New York Medical Repository*,[1] he points out the difference between angina trachealis and putrid sore throat, or, in modern terms, between membranous croup and diphtheria. His observations were founded upon autopsies of the two diseases, and therefore rested on an anatomical basis. It is unfortunate that his views did not attract more attention, and make a more permanent impression than they did. They were corroborated by Dr. Peter Middleton, of New York, who satisfied himself that croup "is totally distinct from the malignant sore throat; it is not of itself of a nature malignant or infectious as the putrid sore throat may often be." These views were put forth nearly ninety years ago, and have been confirmed only within a comparatively recent period. Among those who have studied the natural history of this disease, Dr. John Ware deserves honourable mention. His memoir on the history and diagnosis of croup contributed materially to the accuracy of our knowledge of it and to its correct treatment. His paper was based upon a careful study of the cases which came under his own observation. He was satisfied that membranous croup and

---

[1] New York Medical Repository, vols. xii. and xiv., 1809 and 1811.

inflammatory croup were not different stages of the same disease, but distinct maladies, differing from each other in their character and prognosis, and requiring a different treatment. His reasons for believing in the essential difference of the two diseases are stated in the following moderate language: "The very great preponderance of fatal results in the membranous croup and a similar preponderance of recoveries in the inflammatory, and the evidence which exists that in a few cases of recovery from the former the membrane has been found, and in the few cases on record of death from the latter that a membrane has not been found, afford strong reason for believing that the diseases are essentially different."[1] Dr. Ware regarded the membrane in membranous croup more as a result of a peculiar kind of inflammation than as the essential part of the disease. As to the prognosis in the two forms, he inferred from his observations "that the only form of croup, attended with any considerable danger to life, is that distinguished by the presence of a false membrane in the air passages." To this he added the following remark: "The existence of this membrane in the air passages is in a very large proportion of instances indicated by the existence of a similar membrane in the visible parts of the throat." As far as treatment is concerned, he was satisfied that inflammatory croup gets well sooner by the aid of mild and soothing applications, such as emollient gargles, light diet, opiates, and occasional poulticing externally, than by heroic treatment, such as opening the jugular vein, free leeching, antimonial and other emetics and violent cathartics, with which this disease has been so frequently and unfortunately combated. In like manner it was a fair induction from his cases that membranous croup is more likely to be aggravated than relieved by violent appli-

---

[1] Contributions to the History and Diagnosis of Croup, by John Ware, 1842.

cations. He found that the inhalation of warm vapour, an even temperature, and enforced quiet more frequently led to the resolution of the inflammation and consequent detachment and expulsion of the membrane than the lancet or caustic or other extreme measures. When we consider that these observations were made, and the record of them and deductions from them published more than thirty years ago, and observe how nearly they represent our present knowledge of the history, prognosis, and treatment of croup, we cannot resist the conclusion that Dr. Ware was largely in advance of his time in comprehending the nature of croupal affections, and that his observations on these affections were a valuable contribution to practical medicine. The minute studies of recent German investigators in this direction have substantially confirmed Dr. Ware's earlier views.

Abernethy was in the habit of urging with great earnestness the importance, especially so far as treatment is concerned, of the constitutional origin of local diseases. The late Dr. Horace Green, of New York (who achieved such a large notoriety as a specialist in diseases of the throat), insisted with equal emphasis upon the local origin of constitutional diseases. His treatise on diseases of the air passages might be regarded without injustice as a defence of such a thesis. Its real object was, of course, to present and defend Dr. Green's peculiar views. Although his pathology and therapeutics were severely, and to a large extent, not unfairly criticized, both in America and Europe, yet it is not to be denied that his observations contributed to advance our knowledge of the throat and its maladies. They not only stimulated inquiry, but showed how far local applications could be carried into those regions, and to what extent the tissues would bear cutting, slashing, and burning. Dr. Green was a bold and skilled operator, an heroic therapeutist, and was sometimes charged with magnifying his office. These qualities enabled him to do what others would have shrunk from. We must remember that the laryngoscope, which has revolutionized our notions of the throat as much as Laennec's discovery did our notions of the chest, was not known when Dr. Green was studying and treating the air passages. Since the laryngoscopic mirror has rendered visible parts of the throat that were previously invisible except after dissection on the dead body, and has rendered possible a variety of local applications and operations that would not previously have been ventured upon, it has been ascertained that Dr. Green's attempts to reach and act upon the glottis, epiglottis, larynx, even down to and below the bifurcation of the bronchi, were legitimate. Czermak, Mackenzie, and their disciples have carried the local treatment of the throat and air passages much further than Dr. Green ever attempted, but he deserves the credit of having opened the way into a region which later physicians with better appliances and ingeniously constructed instruments have explored with such success.

Autumnal catarrh, commonly called hay fever, from some supposed, but improbable and unproved connection with hay as its cause, has been recognized as a distinct disease only within a comparatively short period. It has undoubtedly been one of the ills flesh is heir to from time immemorial, but has been confounded with ordinary catarrh, asthma, and the like. Gradually its distinctive features have been made out. Since it has obtained the status of a distinct disease, it is surprising how many people, both in this country and in Europe, have been found to be sufferers from it. Few or none die from it, and the consequent inability to obtain post-mortem information makes our knowledge of its pathology more or less conjectural. Dr. Morrill Wyman, of Cambridge, whose interest in the disease may possibly be heightened by the fact that he is personally one of its victims, has contributed more than any other observer that we are acquainted with, to a correct knowledge of its natural history and treatment. His treatise on autumnal catarrh[1] is a classical one of its kind. He has pointed out the distinctive characters which separate it from other catarrhs, its limited duration, its remissions, intermissions, and whimsical variations, its intractability to the action of drugs, the fact that certain regions are free from it and that migration to these regions relieves the sufferer almost immediately. He has made several experiments with regard to its etiology, which, while they do not demonstrate its cause, indicate the direction of study which will probably lead to the discovery of its cause, and has shown that "the disease has more of a general than local character, and falls especially upon the nervous system." By long and careful observation he has ascertained that the regions of this country which are free from the disease, the places of refuge for the catarrhly afflicted, are the northern side of the White Mountains in New Hampshire; Mount Mansfield, in Vermont, and its immediate neighbourhood; the Adirondacks, in New York; the Ohio and Pennsylvania plateau, including the high range of land in New York from the Catskill Mountains to the western border of the State; the island of Mackinaw; the northern side of the great lakes in Canada; tracts of land beyond the Mississippi, at St. Paul and in Minnesota; the Alleghany Mountains at Oakland, and other elevated points of the same region; the high lands of the interior of Maine; and the whole sea-coast from St. John's quite round to Labrador. It thus appears from Dr. Wyman's observations that the regions of safety for the afflicted are by no means small; and that in this disease climate most effectually supplements the action of drugs.

It is undeniable that during the past century, and particularly during the past fifty years, medical science has made great and satisfactory progress in acquiring an intimate and

---

[1] Autumnal Catarrh (Hay Fever), New York, 1872.

accurate knowledge of the natural history, pathology, and appropriate treatment of diseases of the chest and air passages. When we consider the contributions to this progress made by the American physicians, Bowditch, Gerhard, Pennock, Da Costa, Alonzo Clark, Austin Flint, Green, Ware, Wyman, and others whose labours we have so imperfectly described, and by other physicians whose contributions we have not time to mention, we can point with honest pride to the honourable record of service rendered to the progress of this department of medical science by America.

In our allusion to Dr. Nathan Smith's papers on fevers, we referred to his conjecture, or belief, that typhoid fever could not be broken up by treatment; that it was in fact a self-limited disease. The best observers at that time were beginning to reach that conclusion. Louis' observation of typhoid fever led him to entertain the same notion; Andral's study of typhus led him, so far as that disease is concerned, to the same conviction. Doubtless there were other observers scattered here and there in Europe and America, who had learned to recognize the fact that some diseases were self-limited in their course, but such was not the common view. Dr. Jacob Bigelow, in a paper published in 1822,[1] was the first or among the first to make a clear and distinct statement—a grand generalization from the study and observation of disease—that self-limitation is one of the laws that govern the course of a large number of morbid processes. This paper is not only a statement of the law, but a demonstration of its truth. Dr. Bigelow did not claim absolute originality for his views, for in his paper he says: "I am aware that some of the most distinguished French pathologists of the present day incline to the opinion that many acute diseases, or at least inflammations, are incapable of being shortened in their duration by art. The opposite

[1] Mass. Med. Soc. Comm. vol. iii.

opinion prevails very generally in this country and in England, and it would be premature to consider the question as decided, until it has been submitted more extensively to the test of comparative numerical results." That test has since been applied, and has resulted in confirming the accuracy of Dr. Bigelow's statement. We do not partake of the enthusiasm of a medical friend, who said that he would rather have written Dr. Bigelow's paper on self-limited diseases than to have been the victorious commander at Waterloo. Still the paper was one of those clear and distinct statements of a truth, or rather of a natural law, which, by directing the attention of physicians in this country and elsewhere to a neglected and unrecognized fact, was an admirable contribution to the progress of practical medicine. It has undoubtedly saved a great many lives by preventing useless and violent medication, and has saved many more by turning the attention of practitioners to the support of the system, while disease was passing through its appointed orbit. The observation of every year since the appearance of Dr. Bigelow's paper has lengthened the catalogue of self-limited diseases. Science is beginning to learn that the laws which govern morbid processes are not less immutable than those which control the planets; and that therapeutics to be rational and successful must conform to these laws, and not undertake to neglect or thwart them.

A superficial observation would lead to the belief that delirium tremens could not be an illustration of the law of self-limitation in disease. Formerly, according to the popular and perhaps universal sentiment of the profession, delirium tremens was an affection that required prompt and active interference. In conformity with such a notion, the heaviest batteries of the materia medica were turned upon the unfortunate victims of this malady, and a rapid and unrelenting discharge of drugs kept up upon them. Opium, emetics, assafœtida, warm baths, digitalis, hyoscyamus, valerian, prussic acid, wormwood, spirits, sulphuric ether, hops, borax, and other articles were prescribed, separately or in combination, with extraordinary activity and zeal. Dr. John Ware, of Boston, was not satisfied with the results of such active and indiscriminate fighting. He accordingly determined to study the natural history of the disease. In 1831, he published a paper[1] on delirium tremens, founded exclusively upon a considerable number of cases of it which had occurred under his own observation. This memoir is an original one of marked value, and of special clinical interest. In it the expectant treatment during the paroxysm is highly spoken of, and its result is stated to be a termination of the attack "at a period seldom less than sixty, or more than seventy-two hours, from the commencement of the paroxysm." This result of the expectant treatment as demonstrated by his observations, he compared with the results of other kinds of treatment, as reported by those who have tried them. The inference from the comparison is not in favour of active interference. "I am satisfied, therefore," says Dr. Ware, "that in cases of delirium tremens the patient, so far as the paroxysm alone is concerned, should be left to the resources of his own system, particularly that no attempt should be made to force sleep by any of the remedies which are usually supposed to have that tendency; more particularly that this should not be attempted by the use of opium." Since the introduction of bromide of potassium and chloral hydrate as hypnotics, patients with delirium tremens have been enabled, by the aid of these agents, to pass more comfortably through the paroxysms of the malady, but it is doubtful whether the period of sleeplessness has been curtailed by them. The observation of more than forty years, that have elapsed since the appearance of Dr. Ware's paper, has confirmed the accuracy of his statements, and has also shown that

[1] Transactions of the Mass. Med. Society, Boston, 1831.

delirium tremens is one of the diseases included by the law of self-limitation. Dr. Kuhn, of Philadelphia, treated this malady, nearly a century ago, after the expectant fashion[1] in a novel way, by "confining the patient in a dark cell, and leaving the disease spontaneously to work itself off." After an extensive trial of this method he was satisfied that it yielded a good result. He experimented also with the opium treatment[2] in 1783. The observations of Dr. Ware, which have just been cited, confirmed the earlier results of Dr. Kuhn.

Medical science is largely indebted to Dr. Austin Flint, of New York, an observer whose acquirements, accuracy of observation, and soundness of judgment have justly earned for him a European as well as an American reputation, for ascertaining that an affection so apparently irregular in its course, and so generally supposed to require active treatment as rheumatism, belongs to the class of self-limited diseases. A recent paper[3] of his contains a series of clinical observations on the treatment of acute articular rheumatism. It is not to be forgotten that Oppolzer instituted a similar inquiry some years ago with regard to the same disease. The distinguished German physician felt justified by his observations in asserting that drugs might mitigate the distress, and prevent or relieve some of the complications of rheumatism, but could not shorten its natural termination. Dr. Flint's observations confirm those of Oppolzer, and indicate that the rational treatment of this intractable affection consists in keeping it within its natural orbit, and not in vain efforts to curtail it.

Recent investigations, especially those of Charcot in locomotor ataxia, and those of Weir Mitchell on injuries

---

[1] Phila. Journ. Med. and Phys. Sciences, iii. 242.
[2] N. A. Med. and Surg. Journ., iv. 235.
[3] Am. Journ. Med. Sciences, July, 1863.

of nerves, have disclosed an unexpected relation between certain derangements of the spine and swelling of the joints. More than forty years ago Dr. J. K. Mitchell, of Philadelphia, was led by rheumatic or rather by rheumatoid symptoms, in a case of caries of the spine, to suspect a connection between the medulla spinalis and the supposed rheumatism. He collected a number of cases besides those which came under his own observation, and founded upon them two papers,[1] one of which appeared in 1831, and the other in 1833. His observations were original and valuable. Their author did not follow them to the legitimate conclusions, which later investigations show might have been drawn from them. Nevertheless, as far as he went, he was in advance of his time. Dr. Flint has also recently published a paper on the natural history of acute dysentery, founded upon a series of cases observed and treated by himself. One of the practical conclusions which he drew from the study of these cases is that dysentery "is a self-limited disease, and its duration is but little, if at all, abridged by methods of treatment now and heretofore in vogue."[2]

In March, 1864, Dr. John C. Dalton read a paper[3] before the Academy of Medicine in New York, giving an account of some observations which he had previously made on Trichina Spiralis. In 1869 he supplemented this paper by another one[4] on the same subject. These two papers not only contain an account of what was previously known with regard to this curious parasite, but a number of interesting original observations upon trichinæ taken from trichinous meat, and also upon those taken from man. The two papers are valuable contributions to the natural history of

[1] Amer. Journ. Med. Sciences, May, 1831, and August, 1833.
[2] Ibid., July, 1875.
[3] Transactions of the New York Academy of Medicine, 1864.
[4] Medical Record, N. Y., April 15, 1869.

trichinæ and to the best method of protecting the system from their ravages.

To arrange a series of facts so as to compare them with each other, ascertain their mutual relations, draw from them legitimate deductions, and thus demonstrate some unknown truth, or confirm one previously recognized, is to render as distinct a service to the cause of science as to collect the facts themselves. Indeed, the collection of facts without comprehending their relations to each other and to the whole world of facts, is a barren service.

Dr. Oliver Wendell Holmes, whose brilliant reputation as poet and novelist must not make us forget that he is also physician and anatomist, prepared a paper[1] in 1843 upon the important subject of the contagiousness of puerperal fever, a paper which belongs to the former of the two classes of contributions to medical science that we have just mentioned. The practical point which Dr. Holmes illustrated and proved is that "the disease known as puerperal fever, is so far contagious as to be frequently carried from patient to patient by physicians and nurses." The merit of this paper consists, not only in the collection and arrangement of the evidence that had accumulated upon an important matter, but in the logical and forcible presentation of the argument which the evidence legitimately warranted in favour of the point he maintained. Its value as a contribution to practical medicine is shown not only by the influence it exerted in this country, but also by the fact that Copland and Ramsbotham referred to it in approving terms, and that the Registrar General of England made use of it in his fifth annual report. It is interesting to note, that at the time when Dr. Holmes' paper appeared, two works that were largely, if not almost universally appealed to, as authorities in this country, viz.,

---

[1] This paper was published in the New England Journal of Medicine and Surgery for April, 1843.

Dewees' Treatise on the Diseases of Females, and the *Philadelphia Practice of Midwifery*, by Dr. C. D. Meigs, both taught the non-contagiousness of puerperal fever. At the present time the question may be considered settled in favour of the view which Dr. Holmes deduced from the facts which were then in his possession.

Not long after the appearance of Dr. Holmes' paper Dr. Samuel Kneeland, Jr., published one[1] on the connection between puerperal fever and epidemic erysipelas, in which he maintained that the two diseases are similar. His paper presented the evidence in favour of this view. Within the past year another American physician, Dr. Thomas C. Minor, of Cincinnati, has published a work[2] in which he enters into a careful and elaborate examination of the relations of puerperal fever to erysipelas, based upon the facts obtained from the census of the United States for 1870. Among the conclusions which Dr. Minor felt warranted in drawing from the evidence before him are the following:—

1st. That there is an ultimate connection existing between child-bed fever and erysipelas, and that in any place where erysipelas is found there will be found puerperal fever.

2d. Physicians attending child-bed fever cases and erysipelas at the same time were most unfortunate in their practice

3d. Physicians having large obstetric practices, but who are known to be believers in the close connection of child-bed fever and erysipelas, returned few death certificates from either cause.

4th. Epidemic erysipelas is invariably associated with an outbreak of epidemic child-bed fever, or *vice versâ*. The *London Practitioner* for August, 1875, in a notice of Dr.

---

[1] Am. Journ. Med. Sciences, April, 1846.

[2] Erysipelas and Child-bed Fever, by Thomas C. Minor, M.D., Cincinnati, 1874.

Minor's work, says: "If it be asked what was the bond of the connection between erysipelas and child-bed fever here maintained, the same conclusion is suggested by the American as by the English experience, namely, chiefly the *doctor* and the *nurse*."[1]

Notwithstanding the care with which Asiatic cholera has been studied by competent observers in almost every part of the world, its pathology and treatment have not yet been clearly made out. We are indebted to Dr. William E. Horner, the distinguished anatomist of Philadelphia, for discovering one important fact with regard to it. The origin of the rice-water discharges in that disease had long been an unsolved problem. Dr. Horner first detected the fact that in cholera the whole epithelium is stripped from the small intestines, and that the turbid rice-water dejections, which are so characteristic of this disease, result from this peculiar stripping of the mucous membrane. For an account of Dr. Horner's researches, which were made with his singular patience and accuracy, and which led him to the discovery of this pathological fact, we must refer to his original article.[2] Here we can only call attention to his early recognition of it. His recognition and published record are

---

[1] The different departments of medical science naturally and inevitably run into each other to such an extent that it is impossible to draw a distinct line of demarcation between them. Their boundaries are fluctuating and indeterminate. It might be justly said that these references to puerperal fever belong more properly to the Report on Obstetrics and Gynæcology which will appear hereafter, than to one which is concerned only with practical medicine. On the other hand, erysipelas comes chiefly under the eye of the general practitioner. The mutual relations of erysipelas and puerperal fever may, therefore, be discussed as appropriately under the head of Practical Medicine as under the head of Obstetrics. Neither the general practitioner nor the obstetrician can afford to neglect them.

[2] Am. Journ. Med. Sciences, vol. xxi. page 289.

illustrations of our previous statement, that American journals contain accounts of numerous isolated facts pertaining to the various branches of medicine which show that American physicians have not been idle scientific observers. The same observer instituted in 1827 a series of original and interesting inquiries into the healthy and diseased appearances of the gastro-enteric mucous membrane. He endeavoured to ascertain the healthy condition and appearance of this membrane, its appearance in congestion from the agonies of dying, and its appearance in genuine red inflammation. His conclusions were, that congestion is not an active condition of the part affected, but is most frequently the result of mechanical impediment to the venous circulation.[1] Dr. Horner also instituted an inquiry into the anatomical characters of Infantile Follicular Inflammation of the Gastro-intestinal Mucous Membrane, and into its probable identity with cholera infantum. This paper pointed out very clearly the changes which occur in the follicular apparatus.[2] Dr. Horner's labours in other directions, which have given him such a distinguished place among American anatomists, do not fall within the scope of this essay.

The investigations of Dr. John Neill, of Philadelphia, on the mucous membrane of the stomach, made a quarter of a century ago, were original, and added to our knowledge of the structure of that organ. The results of his investigations were given to the public in a paper[3] entitled "On the Structure of the Mucous Membrane of the Stomach," which may be consulted at the present day with advantage.

The liver has always presented an interesting and difficult field of study to the physiologist, the pathologist, and the practitioner. The problems which it offers to the student are far from being solved at the present day. While there is

[1] Amer. Journ. of Med. Sciences, vol. i. 1827.
[2] Ibid., vol. iii. 1828.   [3] Ibid., Jan. 1851.

an agreement on many and important points among medical scientists, there are many others which are still debated. Some of the ablest living physiologists and histologists, like Claude Bernard, Ch. Robin, Kölliker, Schiff, and others, have been led by their investigations to entertain and defend different, and, sometimes, opposing views of the intimate structure and functions of the liver. American physicians and physiologists have not been mere spectators of these efforts to disentangle and clear up such knotty questions. Dr. Leidy's paper on the comparative structure of the liver[1] is the most exact and complete essay in the department of microscopic anatomy which has appeared in any American medical journal, and is a most valuable contribution to our knowledge of the liver.

To this we may add the researches of Dr. Austin Flint, Jr.,[2] of New York, upon cholesterine, which have thrown a good deal of light upon one of the obscure functions of the liver. According to him, says Küss,[3] "the excrementitial product formed by the disassimilation of the brain and of the nerves, at the expense of protagon, is represented by cholesterine, separated from the blood by means of the liver, and then thrown into the intestinal canal. This view is based upon a number of experiments which show, moreover, that the excretion of cholesterine is in direct ratio with the nervous activity. The common expression, to feel bilious, seems justified by one of the elements of the bile, viz., cholesterine." The connection between derangements of the liver and disturbance of the functions of the brain has long been clinically recognized. Whatever explains the mechanism of this connection, is as much a contribution to practical medicine as to physiology. Professor John C. Dalton, of New

[1] Amer. Journ. of Med. Sciences, Jan. 1848.
[2] Ibid., Oct. 1862.
[3] Lectures on Physiology, by Professor Küss, translated by Robert Amory, M.D., p. 27.

York, has rendered efficient service in this direction by his efforts to explain the glycogenic function of the liver. Schiff and Pavy maintain that the sugar found in the liver is a post-mortem product. Dr. Dalton, whose experiments were conducted, to say the least, with as much care, ingenuity, and rapidity as those performed by Schiff, Pavy, or Bernard, demonstrated the presence of sugar in the living liver. The practical relation of his experiments and their result to the question of diabetes is obvious.

Our present knowledge of gastric digestion is largely due to the opportunities, which gastric fistulæ have afforded physiologists for the inspection of the living stomach, or more exactly, of the stomach at work. Medical science owes a debt of gratitude to Dr. William Beaumont, surgeon in the U. S. Army, for leading the way in this method of experiment and observation. The subject of his experiments was Alexis St. Martin, a French Canadian voyageur, who was wounded in 1822, in such a way as to produce a permanent gastric fistula. Fortunately Dr. Beaumont was able to keep St. Martin under his observation for a long time. By means of the fistula he made a series of extended, careful, and valuable experiments upon the digestibility of different articles of food and drink, and noted the behaviour of the stomach in a state of quiescence and in one of activity. His experiments and the inferences which he drew from them are so well known that it is unnecessary to describe them here. They are not only valuable in themselves, but opened the way to a method of investigation, which, both in this country and in Europe, has yielded in the hands of various physiologists important results to practical medicine.[1]

Dr. J. J. Woodward, now in Washington, D. C., published

---

[1] The Physiology of Digestion, with Experiments on Gastric Juice, by William Beaumont, M.D., U. S. A. The first edition of this work was published in 1833.

in 1864 a work on the *Chief Camp Diseases of the United States Armies as Observed during the Present War*. It is a practical contribution of great value to military medicine, and can be studied with profit by physicians in civil practice. Its account of camp diarrhœa and malarial fever are of especial interest.

No department of medical science has been studied with greater earnestness than that of the nervous system. Its importance justifies the labour and time expended upon it. Among American physicians and physiologists who have endeavoured to unravel its intricacies, Dr. S. Weir Mitchell, of Philadelphia, is *facile princeps*. He has done much valuable work in this direction, to which we can only briefly refer.

During our late civil war Dr. William A. Hammond, of New York, himself eminent as a neuro-physiologist and neuro-pathologist, established, at the time he was Surgeon-General of the United States Army, a hospital for nervous diseases, and invited Dr. Mitchell to take charge of it. Drs. Morehouse and Keen were associated with Dr. Mitchell in the management of the hospital. The experience acquired in this hospital by the gentlemen in charge of it led to the publication of a number of communications by them on nervous affections. One of the most important of these was entitled "Gunshot Wounds and other Injuries of Nerves," published in 1864. In the language of Dr. Mitchell, "this volume describes at length all the primary and secondary results of nerve wounds, and especially many hitherto undescribed lesions of nutrition, as well as a novel form of burning pain previously unknown, as a consequence of gunshot wounds. There are also full details of treatment, and a report of thirty-one cases of nerve lesions." With regard to this book, the *Edinburgh Medical Journal* says that it is valuable to practical surgeons, from the many details of treatment which it contains, and that it is "specially interesting to physiologists and neuro-pathologists, from the extreme care with which the cases appear to have been taken, and the exactness and minuteness of the descriptions of the effects of the injuries on motion and sensation." "The glossy skin," previously noticed by Paget, is here described in detail, and shown in many cases to be connected with the peculiar burning pain that is noticed. The same observers put forth a paper on Reflex Paralysis in 1864. In this paper a novel theory of "shock" from injuries is set forth, and cases related where a ball-wound of one limb caused paralysis of remote parts of the body.

The monograph on gunshot wounds was supplemented by Dr. Mitchell in 1871, by a memoir on "The Diseases of Nerves resulting from Tying." This was published in the medical volume of the Reports of the U. S. Sanitary Commission. In 1872 Dr. Mitchell published a work upon Injuries of Nerves and their Consequences, which he dedicated to Dr. Wm. A. Hammond, "whose liberal views," says Dr. Mitchell, "created the special hospital which furnished the chief experience of this volume." The work was chiefly based on the author's own observation. The *British and Foreign Medico-Chirurgical Review*, in a notice of this treatise, says it is "the first complete treatise on the subject the English language has been in possession of," and adds, the volume is "written not only up to the present time, but in many respects far in advance of it," to be referred to now and in the future " with the utmost confidence and satisfaction." In 1874 Dr. Mitchell published a paper on post-paralytic chorea,[1] in which he pointed out the fact that organic palsies, especially hemiplegia, "are occasionally followed by hemichorea, or a still more limited local development of that disorder." In other words, his paper shows

---

[1] American Journal of the Medical Sciences, vol. lxviii. p. 342, Oct. 1874.

that "as there is a post-choreal paralysis, so, also, is there a post-paralytic chorea." Our space forbids our pointing out the amount of original matter and suggestions which these various books and papers contain. As a whole they form the most valuable contributions to neurology and medicine in general which this country has produced. They are admirable as to style, logic, and ideas, and are full of suggestive hints and generalizations.

Any account of American contributions to neurological science and therapeutics, would be incomplete without a reference to the labours of Dr. William A. Hammond in that direction. His investigations upon the physiological action of remedies will be referred to in another place.

In his treatise on sleep[1] he has added materially to our knowledge of the physiology of that mysterious condition, and to the therapeutics of insomnia. So far as priority of discovery is concerned, the credit of ascertaining that sleep is due to a partial anæmia of the brain belongs to Drs. Durham and Fleming. Dr. Hammond, before he had heard of Durham's experiments, made similar ones, and arrived at similar results. His treatise, however, not only gives an account of his own original experiments upon the state of the intra-cranial circulation during sleep, but presents the whole subject of sleep and its derangements in a clear and satisfactory manner.

A full account of Dr. Hammond's contributions to neurological science may be found in his recent work on Diseases of the Nervous System. Of this treatise the author says in his preface: "One feature I may, however, with justice claim for this work, and that is that it rests to a great extent on my own observation and experience, and is therefore no mere compilation. The reader will readily perceive that I have views of my own on every disease considered, and that I have

---

[1] Sleep and its Derangements, by William A. Hammond, M.D.

not hesitated to express them." The size of the work forbids our attempting to analyze it here. For any accurate notion of Dr. Hammond's peculiar views, and original observations, we must refer those interested in the matter to the work itself. We desire, however, to call attention to the account which it gives of athetosis, a disease first recognized and, we believe, first described by Dr. Hammond. His description of this rare affection is illustrated by two cases of it, which have come to his knowledge.

Electro-physiology, and electro-therapeutics for the last twenty-five years, and especially since the appearance of the treatise of Duchenne of Boulogne, upon those subjects, have attracted a great deal of attention. American as well as European observers have been busy with efforts to discover the relations of electricity to the nervous system. By far the most important contribution made by any American observer to this subject, is the treatise[1] of Dr. Charles E. Morgan. Unfortunately the author died before the work went to press. It was published under the editorial care of Dr. William A. Hammond, who thereby bore unequivocal testimony to its value. We learn, moreover, that so high an authority as Professor Rosenthal would gladly have undertaken the revision and editorship of this work, not only as a proof of his esteem for its writer, but also from his conviction of its eminent scientific value. An obscurity of style due partly to a lack of personal revision, and partly to German methods of expression, which the author's long residence and study in Germany had led him into, pervade the book. Whoever masters his style and gets at his thought will agree with the editor "that there is nothing in the English language which at all approaches it as regards the scientific treatment of the whole subject of electricity." It

---

[1] Electro-physiology and Therapeutics, Charles E. Morgan, A. B., M.D., New York, 1868, pp. 714.

is mainly physiological, only about twenty-five pages being devoted to the therapeutics of the subject. The character of the results at which he arrived, and the stamp of his mind, may be derived from the closing paragraph of the book. "Such are the definite scientific applications of electricity to medical purposes; of the many others it need only be said that they are either based on incorrect theory or diagnosis of disease, or an imperfect or incorrect knowledge of electro-physiology; although I do not deny that future researches may enable us to do more, far more than has hitherto been done in this direction."

The New York Society of Neurology and Electrology recently appointed a committee consisting of Prof. John C. Dalton, Dr. George N. Beard, and three others to examine and report upon the existence and localization of motor centres in the cerebral convolutions. The committee made a number of carefully conducted and ingenious experiments upon dogs.[1] The results at which the committee arrived confirmed the most important of those obtained by Hitzig and others who have followed him in this line of experiment. Although these and similar investigations are purely physiological in their character, yet they have such an obvious bearing upon diseases and treatment of the nervous system that they really belong to practical medicine.

The drugs of the Materia Medica, which are fortunately no longer regarded as the only or chief agents by which disease is prevented or combated, still justly hold an important though secondary place in the armamenta medicorum. The contributions of America to this department of practical medicine during the past century have been numerous and valuable. Our space permits a reference to only a few of them. As South America does not come within the limits of our survey, we are prevented from referring to cinchona

[1] New York Medical Journal, March, 1875.

and its alkaloids, a contribution to the resources of medical art, American in its origin, without which the modern practice of medicine would be sadly crippled. Excluding this and all other South American medicinal products from our consideration, let us glance at what the United States has contributed in the past century. As we shall have occasion to see by and by, it has led the way in the introduction of one class of agents whose value cannot be over-estimated.

Contributions to materia medica are of two classes. The first class comprises new, or previously unknown agents, whether vegetable or mineral in their origin, as veratrum viride or wild cherry, and also new chemical combinations, as chloroform or chloral. The second class comprises researches, either clinical or physiological, into the action of medicines, by which their therapeutical power and limits are determined. This class of course includes experiments by vivisections or otherwise on animals and various sorts of chemical analyses.

Let us glance for a moment at the first of these classes. For two or three hundred years previous to the beginning of the present century, there was a popular notion floating about in the community, especially in Germany and parts of France, to which physicians gave very little credence, that ergot was an oxytocic. It was commonly known in Germany by the name of *mutterkorn*, and in this country, as well as in Europe, was sometimes called *pulvis parturifaciens*, names that indicate the popular notion of its power. Notwithstanding the efforts of a distinguished French accoucheur, Desgranges, who recognized its value more than a century ago, and endeavoured to bring it into use, it was forgotten or not accepted by the faculty. Dr. John Stearns, of Saratoga County, New York, in a memoir[1] published in 1808, again called attention to ergot as a remedy for quick-

[1] New York Medical Repository, 1808, vol. xi. p. 303.

ening childbirth. The paper gives an admirable account of the article it describes, and the profession since its time have acquired very little additional information with regard to it, for Dr. Stearns not only recognized its action upon the uterus, but its constringing power over the small blood-vessels, through the intervention of the nervous system. Soon after the appearance of Dr. Stearns' paper other observers confirmed his statements. Dr. Oliver Prescot published in 1814 a paper,[1] giving an account of the natural history and medical effects of secale cornutum. This paper though a less valuable contribution to medical science, than that of Dr. Stearns, had merit enough to be honoured by a French translation, and an introduction into the *Dictionnaire des Sciences Médicales.* The medical profession were now fully aroused to the value of ergot. The use of it spread rapidly over this country, and it was not long before European physicians recognized its virtue. It was established in the place it now holds as one of the important articles of the materia medica. American medical science may fairly claim the merit of restoring to therapeutics an agent, whose virtues Europe had failed to recognize.

We have the authority of the United States Dispensatory for the statement, that "chloroform was discovered by Mr. Samuel Guthrie, of Sackett's Harbour, N. Y., in 1831. At about the same time it was also discovered by Soubeiran in France, and Liebig in Germany." Though the priority of discovery belongs to the American chemist, yet it is evident that the discovery was made by each of the three observers independently of each other; it is also evident that none of them had any notion of the anæsthetic virtue of chloroform to which we shall refer further on. In con-

---

[1] A Dissertation on the Natural History and Medical Effects of Secale Cornutum or Ergot, by Oliver Prescot, Medical and Physical Journal, 1814, vol. xxxii. p. 90.

nection with the importance that chloroform afterwards attained, it is interesting to recall the language which Mr. Daniel B. Smith, of Philadelphia, used with regard to it in 1832. "The action of this ether" (meaning chloroform) "on the living system is interesting, and may hereafter render it an object of importance in commerce. Its flavor is delicious, and its intoxicating qualities equal to or surpassing those of alcohol. It is a strong diffusible stimulus, similar to the hydrated ether, but more grateful to the taste."[1]

Dr. Stillé, in his *Therapeutics and Materia Medica*, makes the statement that the American Indians were acquainted with some of the virtues of podophyllum. At any rate it was for a long time popularly known and used as a cathartic in this country before physicians employed it. Dr. Jacob Bigelow accurately described both the plant and its medicinal properties more than forty years ago. It did not come into general use, however, until its active principle, known as podophyllin, or more exactly resina podophylli, had been extracted. It is now freely used both in this country and in Europe, and cholagogue as well as cathartic properties are attributed to it. Although its virtues have been exaggerated, as have those of leptandrin and gelsemium, yet all of them are valuable additions to the materia medica.

The wild cherry, or prunus virginiana, is another contribution from the flora of America that deserves honourable mention. Its tonic and sedative properties were recognized more than fifty years ago by Dr. John Eberle, whose *Therapeutics and Materia Medica* introduced to the acquaintance of physicians a number of articles, previously unknown or little known, derived from the Americam vegetable kingdom. Dr. Eberle's experiments made upon himself with an infusion of wild cherry, by which he demonstrated its sedative influence upon the heart, deserve to be remembered not only

---

[1] Journal of the Phila. Coll. of Pharmacy, iv. p. 118.

on account of their intrinsic value, but because they show a recognition by him, at that early period, of the importance of making the physiological action of drugs the guide to their therapeutical employment. Dr. George B. Wood, of Philadelphia, one of the most accomplished of American physicians, has pointed out in his *Therapeutics and Materia Medica* the value of wild cherry in phthisis. Some of the foreign journals have also recorded observations in confirmation of these statements.

In 1850, Dr. W. C. Norwood, of South Carolina, proclaimed[1] in somewhat extravagant language the sedative virtues of veratrum viride. Dr. Tully, of New Haven, and other American physicians had previously employed it, but the attention of the profession was not generally directed to it until Dr. Norwood so ardently advocated its employment. Since that time, its active principles, viridia and veratroidia, have been isolated, and they, and the plant from which they are derived, have been subjected to a careful examination, so as to ascertain their physiological action. European and American physicians and chemists have interested themselves in this inquiry. The experiments[2] of Dr. H. C. Wood, Jr., of Philadelphia, are among the most valuable that have been made, and are satisfactory and conclusive. While they do not sustain the extravagant claims of Dr. Norwood, they demonstrate that veratrum viride deserves a place in the materia medica.

As we have already intimated, we cannot undertake to give an accurate catalogue of the numerous articles that have been introduced into the materia medica from the American vegetable kingdom. We can barely refer to such articles as geranium maculatum, whose astringent virtues have a

[1] Southern Med. and Surg. Journal, June, 1850.
[2] Amer. Journ. Med. Sci., 1870, and Philadelphia Medical Times, vols. ii. and iii.

local reputation; sanguinaria canadensis, which posseses emetic and expectorant properties; spigelia marilandica, whose anthelmintic virtues were described a century ago by Drs. Garden and Chalmers, of South Carolina; apocynum cannabinum, an emetic and cathartic, whose vulgar name of Indian hemp has led some practitioners to mistake it for a very different article, the Indian hemp of India; senega or seneka root (introduced by Dr. Tennant, of Virginia, in 1731), whose valuable expectorant properties are recognized in Europe and in this country; serpentaria or Virginia snakeroot, used by the American aborigines as a remedy for snake-bites, and considered by physicians of the present day to have stimulant, tonic, and other properties; eupatorium perfoliatum, which, under the less learned name of thorough-wort, is largely used in domestic practice; lobelia or Indian tobacco, an agent of undoubted activity as an emetic, a sedative, and an expectorant, largely used by a notorious charlatan and his disciples, and which possesses a value that gives it a place in our modern materia medica; gillenia, whose virtues as an emetic would enable it to replace ipecacuanha, if the latter could not be easily obtained; sassafras, sabbatia, gaultheria, and a variety of other plants. These, and other articles that might be mentioned, are additions of greater or less importance to the practitioner's list of remedies. Some of them are only known locally; others have travelled beyond the sea, and enjoy a transatlantic reputation. They are mentioned not only on account of their intrinsic merits, but to show that American physicians have not neglected to explore their native forests and fields, with the hope of enriching the materia medica of the world.

Our account of American contributions to materia medica would be imperfect without a reference to the works of Dr. Jacob Bigelow on medical botany. In 1814 he published an octavo volume upon the plants of Boston and its environs. A few years later, he published his American Medical Botany in three volumes, illustrated. This was a contribution to medical science of the highest order. Its descriptions are accurate, concise, and complete. The fifty years that have elapsed since its appearance have taken very little from, and added very little to his account of the plants he described or of their medicinal virtues. It still maintains its place as an authority upon the subjects of which it treats. In this connection the Medical Botany of Dr. W. P. C. Barton should be mentioned. It covers a different ground from that of Dr. Bigelow, but like his is a valuable addition to medical science.

Let us now pass to the second class of contributions to materia medica, viz., clinical, or physiological researches into the action of medicines upon the human system. American physicians and physiologists have begun to cultivate this delicate, difficult, and important field of inquiry. Prominent among the explorers of this region is Dr. William A. Hammond, of New York. His researches[1] upon the physiological effects of alcohol and tobacco upon the human system, upon albumen, starch, and gum, upon the excretion of phosphoric acid, and upon certain vegetable diuretics have added to our knowledge of these articles. Many of the experiments upon which these papers are founded were made upon himself. His paper upon alcohol appeared nearly fifteen years ago. It was limited chiefly to the action of alcohol upon metamorphosis of tissue. Since its appearance, the well-known, laborious, and entensive researches, in the same direction, by Lallemand, Perrin and Duroy, Anstie, Parkes, Binz, Duchek, and others, have greatly increased our knowledge of the physiological action of alcohol, but have not substantially shaken the conclusions of Dr. Hammond. The diuretics whose action he investigated were squill, juniper, digitalis, and colchicum. His object

---

[1] Physiological Memoirs by William A. Hammond, M.D.

was to ascertain their influence over the quantity of the urine, its specific gravity, and the amount of its solid organic and inorganic constituents. His results explain and confirm the conclusions with regard to the therapeutical value of these drugs that physicians have generally held.

The experiments of Dr. Hammond as to the physiological action of diuretics upon healthy adults, were supplemented by a series of clinical experiments[1] with diuretics by Dr. Austin Flint, of New York. Dr. Flint experimented with squill, digitalis, nitrate of potassium, iodide of potassium, acetate of potassium, colchicum, and juniper. He gave these articles separately and in combination. His conclusions with regard to their action upon the solid and liquid constituents of the urine, substantially confirm those of Dr. Hammond. Dr. Flint modestly calls these researches "a small contribution to the study of the effects of diuretic remedies;" they are not only valuable in themselves, but at the time they were made, fifteen years ago, had an especial value as indicating the proper method of the clinical observation of remedies.

The researches of Dr. H. C. Wood, Jr., upon the physiological action of drugs are admirable illustrations of the modern method of ascertaining their action. We have already referred to his study of the action of veratrum viride and its alkaloids. He has made several other similar contributions, which are embodied in his recent work, *Therapeutics, Materia Medica, and Toxicology*. This work is an original contribution to practical medicine. It not only presents a condensed statement of the author's investigations, but of the best European investigations upon the physiological action of medicines.

In this connection, it should be mentioned, that the recent experiments of Dr. Robert Amory, of Boston, and of Prof. H. P. Bowditch, of the same city, upon the absorption and

---

[1] American Med. Monthly, New York, Oct. 1860.

elimination of the bromide of potassium and its kindred salts, have enlarged our knowledge of the action of those remedies in this direction. Dr. J. H. Bill, U. S. A., made a series of experimental researches into the action and therapeutic value of the same article, which he published[1] in 1868. The experiments were made on man. They were carefully conducted, and led him to the conclusion that "bromide of potassium, in its legitimate action, is an anæsthetic to the nerves of the mucous membranes and a depressor of their action."

Dr. S. Weir Mitchell, assisted by Drs. Keen and Morehouse, made some admirable observations and experiments,[2] at the United States Army Hospital for injuries and diseases of the nervous system, upon the antagonism of atropia and morphia. They also made an examination into the power of conia, daturia, atropia, and morphia to destroy neuralgic pain. Their observations led them to the conclusion that morphia and atropia act as mutual antagonists in certain of their effects. The paper is marked by the care, accuracy, and completeness that characterize all of Dr. Mitchell's researches. It deserves to be read in connection with Dr. Fraser's exhaustive researches in the same direction.

An American, travelling in Hungary not long ago, attended some sort of a public meeting in one of the large towns of that region. One of the speakers had occasion to allude, in the course of his remarks, to Boston in the United States. He referred to it as a place well known to the audience, and distinguished not as the cradle of the American revolution, not for its commerce, not for its literature, not for its statesmen, its authors, its poets, and its theologians, not for its manufactures, not for its common school system, but as the place where anæsthesia by means of sulphuric ether

---

[1] Amer. Journ. Med. Sciences, July, 1868.
[2] Ibid., July, 1865.

was discovered. This discovery has rendered the name of Boston familiar to the dwellers on the Danube and the Caspian. Such a fact may not be gratifying to the vanity of Bostonians, but it testifies to the universal recognition of the inestimable value of artificial anæsthesia. In every part of the civilized world, and wherever, in those regions called uncivilized, the pioneers of civilization have penetrated, in Japan, in China, in the islands of the sea, the power to produce anæsthesia, which ether first revealed, is acknowledged and blessed. The knowledge of it is so universal, and the blessings which attended its use are so constant, that we are sometimes apt to think as little of its existence and power as we do of the presence and power of light. It is impossible to estimate or form any adequate conception of the amount of human suffering which anæsthetics have relieved and prevented. To their discovery the human race owes the blessing that no pain follows the course of the surgeon's knife into any living tissue; that the accoucheur can, when necessary, alleviate or abolish the agonies of travail; that sleep can be procured in spite of any agony; and that at the word of the physician any sufferer may be rendered unconscious of torture. Such a power, which John Baptista Porta had strangely prophesied centuries ago, which mesmerism hinted at, which mystics have now and then proclaimed, but which the world never dared to expect, was shown to exist and to be capable of safe and easy application by the use of sulphuric ether at the Massachusetts General Hospital in 1846 It was, perhaps, the greatest contribution to practical medicine that the world has ever received. Of itself, it is enough to render American medical science honoured and memorable.

As soon as it was generally known that the inhalation of the vapour of sulphuric ether would produce insensibility, experiments were made with various substances by physicians in this country and in Europe, who hoped to discover other agents, the inhalation of whose vapour would produce anæsthesia as well as ether, or, perhaps, better than that article. The most distinguished of these experimenters, Sir James Y. Simpson, of Edinburgh, tried chloroform, a substance which was previously regarded chiefly as a chemical curiosity. The experiment of Sir James, made upon himself, disclosed the fact that chloroform was an anæsthetic of great power. The inhalation of its vapour acted rather more rapidly than the inhalation of ether, was less disagreeable to the patient, and in a very small quantity produced profound anæsthesia. The knowledge of the discovery of this new anæsthetic spread rapidly over Europe and this country. The ease of its application, the profound insensibility which it produced, its freedom from any unpleasant odour, and the fact that it was discovered in Europe, while ether came from the wilds of America, and other circumstances, led to its adoption almost universally throughout Europe as an anæsthetic in preference to ether. It was also used very largely in this country, but not as exclusively as on the other side of the Atlantic. The experience of a quarter of a century has shown that the inhalation of chloroform is fatal in a certain proportion of cases, while the inhalation of ether is comparatively innocuous.

American medical science has not only contributed to practical medicine the discovery of artificial anæsthesia, but insists that the anæsthetic, which it first presented to the world, is still the best that is known to science; it insists upon its demonstration of the fact that pure concentrated sulphuric ether is preferable, as an anæsthetic, to chloroform; and, as a logical conclusion from this, it insists that the persistent use of chloroform by European physicians is a grave and serious error. This is not the place, even if we had the time, to discuss the comparative merits of ether and chloroform. Physiological experiments and clinical experience are both in favour of ether, and against chloroform. The prolonged inhalation of ether by man affects, first, the cerebrum; second, the sensory centres of the cord; third, the motor centres of the cord; fourth, the sensory; and fifth, the motor centres of the medulla oblongata. When ether kills, it does so by producing asphyxia, leaving the pulsations of the heart to warn the surgeon of the approach of danger—warnings which only the most reckless carelessness can fail to notice. Chloroform, like ether, affects chiefly the brain and spinal centres; but its action upon the sensory and motor centres is more rapid than that of ether. Upon the heart it produces a steadily depressing influence. When it kills, it does so by cardiac paralysis, acting directly upon the heart-muscle, and not by asphyxia; consequently there is no warning of impending death, and the greatest carefulness cannot avert the fatal issue. When ether kills, death is due to the carelessness of the operator; when chloroform kills, death is due to the rashness, wilfulness, or ignorance of the operator in the selection of his anæsthetic. Dr. H. C. Wood, Jr., whose contributions have been previously referred to, says :—

"As an anæsthetic, chloroform possesses the advantages of quickness and pleasantness of operation, smallness of dose, and cheapness. These advantages are, however, so outbalanced by the dangers which attend its use, that its employment under ordinary circumstances is unjustifiable. It kills without warning, so suddenly that no forethought, or skill, or care can guard against the fatal result. It kills the robust, the weak, the well, the diseased, alike; and the previous safe passage through one or more inhalations is no guarantee against its lethal action. Statistics seem to indicate a mortality of about one in three thousand inhalations; and hundreds of utterly unnecessary deaths have now been produced by the extraordinary persistence in its use by a portion of the profession. It ought, therefore, never to be employed except under special circumstances, as in some cases of puerperal eclampsia, when a speedy action is desired, or in the field during war time, where the bulkier anæsthetics cannot be transported."[1]

These pages have made no reference to the contributions which American medical science has made during the past century to surgery, to obstetrics, including gynæcology, or to medical literature. An account of these contributions, and of the work of American physiologists, will be presented in future essays. The limited portion of the field, however, which we have surveyed has yielded enough of interest and importance to justify an honest pride on the part of American physicians in the work they have accomplished, and to give assurance of continued and zealous labour in the future.

The blank pages of the book, containing all the secrets of medicine, which Boerhaave bequeathed to the future, were prophetic of the work which medical science was destined to accomplish. The science of his age could inscribe only a single sentence upon a single page. The present century, whose closing hours the nation celebrates, has filled two or three additional pages with the secrets it has discovered, calling them vaccination, anæsthesia, and preventive medicine. It now transmits the volume to the coming ages, confident that each succeeding century will make new discoveries, till all of Nature's secrets are discovered, and then the title of the book shall be the just index of its contents.

The discovery of artificial anæsthesia was an event of such transcendent importance that it becomes necessary to give a complete account of it in this connection. As soon as its value was established a number of individuals claimed the honour of its discovery. The controversy which the various

[1] Treatise on Therapeutics, etc., by H. C. Wood, Jr., M.D., p. 251.

claimants and their partisans have carried on has been prolonged and sometimes bitter. Most of those who were familiar with the way in which the discovery was introduced to the public and acquainted with the claimants to it, and in a position to form an impartial and correct opinion of the value of their claims, are no longer living. Of the surgeons of the Massachusetts General Hospital, who were present when the first operation under ether, the experimentum crucis of the new discovery, was performed, only one is now living. Fortunately that one, Dr. Henry J. Bigelow, the distinguished and accomplished professor of surgery in Harvard University, was not only present when the first operation was performed, but was personally acquainted with most of the steps in the early progress of the discovery, with the claimants to the honour of it, and with all of importance that appertains to the history of it. He did more than any other living person to bring it before the medical public of this country and of Europe, to assert its real value, and to point out the best methods of utilizing it. A quarter of a century has elapsed since its discovery. This period is long enough for the heat of partisanship to cool, and to afford an opportunity for an impartial statement, by an impartial observer, to be heard with calmness and interest. Moreover, the centennial anniversary of the nation's existence is an auspicious moment for putting on record the testimony of an intelligent and disinterested witness to the discovery of an agent in which the whole human race are interested. With the hope of obtaining from Dr. Bigelow a history of the discovery of anæsthesia, the following note was addressed to him:—

"Dr. HENRY J. BIGELOW, *Professor of Surgery in Harvard University.*

DEAR SIR: I am preparing a report on the progress of practical medicine in this country for the past century. In such an essay a notice of America's greatest contribution to medicine, modern anæsthesia, is indispensable. If you have the time and are willing to prepare a history of its discovery, with which you are so familiar, you will not only confer a favour upon all interested in it, but will put on record an authentic account of the discovery, by one who was an eye witness and actor in its early scenes, and to whose statements, on account of their disinterestedness, great value is attached."

To this letter the following paper was received in reply.

A HISTORY

OF THE

DISCOVERY OF MODERN ANÆSTHESIA.

BY

HENRY J. BIGELOW, M.D.,
PROFESSOR OF SURGERY IN HARVARD UNIVERSITY.

A HISTORY

OF THE

DISCOVERY OF MODERN ANÆSTHESIA.

MY DEAR SIR: A quarter of a century ago, your simple proposition would have re-awakened a discussion which had already exhausted the subject. Even so long ago as 1848, I thought it discreet to preface a paper upon the abstract question of discovery, as decided by historical precedent, with the disavowal of an intention to "dig up the well-worn hatchet of the ether controversy." But I see no objection to a review now—final, so far as I am concerned—of the occurrences you refer to, especially if I offer no opinion without its reason.

The singular persistence of the controversy was due to a variety of causes. People differ in their views about what constitutes a discovery or a discoverer. A voluminous mass of sworn testimony availed little in those days, for want of some machinery to reach and fix a decision based upon it. One of the contestants, at least, felt this, and vainly urged a court of justice.[1] Preponderating opinions and evidence were laboriously and repeatedly brought to the surface by Congressional committees, and by other bodies and committees of those most familiar with the circumstances; but, for the lack of a tribunal accustomed to estimate the weight

[1] See Congressional Debates for the sessions of 1853 and 1854.

and quality of scientific evidence, not to say evidence of fact, no absolute decision was reached. The result was, that every new discussion began and ended like the preceding, and to as little purpose.

When the discovery was announced (October, 1846), the circumstances were few and recent, and the details of its progress were known. But when history was obscured, when another State, and even another Nation, had set up each its own discoverer, and readers were perplexed with volumes of new reports and new testimony, it became less easy to sift the evidence. Claims, till then distinct, overlaid each other. Each alleged inventor, with his partisans, aimed to secure the whole honour. Opinions were pronounced by people who knew little of the facts. Attempts were even made, in more than one instance, to forestall or manufacture public opinion, by procuring in promiscuous medical assemblies and legislatures sudden votes designating some discoverer by name, with a view to influence the erection of statues. In a scientific view such votes are not worth the paper that records them; but it cannot be doubted that in a free country every citizen has the inherent right, of which the late Lord Timothy Dexter so liberally availed himself, to erect in his front yard a statue with an inscription.

The claimants to the discovery are three—Dr. Wells, Dr. Jackson, and Dr. Morton. In discussing their claims, we cannot overlook the fact that the discovery was equally possible to either of them, or indeed to any moderately ingenious person whose attention should have been directed to the subject. This greatest single step forward in the history of medicine, like the division of the Roman printing-block, was a very small advance in strictly scientific knowledge. Facts of insensibility to pain, produced both by gas and ether, were already known to the world. Art needed only an extension of their application; and so far as art or science was involved, either Wells, Jackson, or Morton was competent to the work.

This simple statement comprehends certain points of vital importance. The first essential requisite of modern anæsthesia is, that it shall be always attainable, and, when attained, complete. A second requisite is, that the insensibility shall be safe. The discovery embraced the threefold and essential novelties, that it is, under proper guidance, *complete, inevitable*, and *safe*, and not, like all previous stupefaction, partial, occasional, or dangerous. If only partial or occasional, or if dangerous, neither the patient, the dentist, nor the surgeon would consider it of value. Even so late as a year after the discovery, many surgeons, and, extraordinary as it may now appear, some hospitals, absolutely declined to use the new means of producing insensibility, on the ground that it was attended with danger.

Readers of the present day may not remember how surprisingly far previous knowledge of anæsthesia had extended. Although there has been a want of discernment in attempts to point out precisely what the new advance upon previous knowledge was, one great difficulty has been, that this advance was so small, in a strictly scientific view, that it was not easy to measure it. This difficulty was enhanced by the magnitude of the spoil, whether mere honour, or, as beyond all question it should have been, emolument.

A rapid review of the history of early anæsthesia will bring us to the period with which we are especially concerned.

In the anæsthetic state, the action upon the brain may be a primary one, as by its compression, or by hypnotism—or secondary, as by narcotic and inebriating agents absorbed into the blood, from the lungs, the digestive tube, the skin, or other tissues. A few extracts, abridged from the familiar literature of the subject, will show that surgical anæsthesia in these various forms has been long known and longer sought.

The use of poppy, henbane, mandragora, hemp, etc., to

deaden the pain of execution and of surgery, may be traced to a remote antiquity. Herodotus ascribes to the Scythians the use of a vapour of hempseed to produce drunkenness by inhalation. In China, Hoa-tho, in the year 220, administered hashish (*Mayo*) and performed wholly painless amputations; the patient recovering after a number of days. Hashish, described by Herodotus twenty-three centuries ago, is the active agent of the modern *Bhang* of India.

Pliny, who perished A.D. 79, says of mandragora: "It is drunk before cuttings and puncturings, lest they should be felt." Dioscorides gives an elaborate method of preparing mandragora to produce anæsthesia ($\pi o \iota \varepsilon \iota \nu$ $\dot{\alpha} \nu \alpha \iota \sigma \theta \eta \sigma \iota \alpha \nu$) in those who are to be cut or cauterized—" or sawed," adds Dodoneus, and who in consequence "do not feel pain." Half an ounce, with wine, says Apuleins, a century later, enables a patient to sleep during amputation "without sensation." "Eruca" was used by criminals about to undergo the lash. "Memphitis," a "stone," so "paralyzed parts to be cut or cauterized that they felt no pain."

Theodoric, about the year 1298, gives elaborate directions how to prepare a "*spongia somnifera*," by boiling it dry in numerous strong narcotics, and afterwards moistening it for inhalation before operations. In 1832, M. Dauriol, near Toulouse, cites five cases of insensibility during surgical operations, induced by him with the aid of a similar "sponge" It is said that persons unpacking opium have fallen suddenly.

In 1532, Canappe described the inhaling-sponge of Theodoric, and at the same time warned surgeons against giving opium (*à boire*), which "sometimes kills." But in later years, and until ether was introduced, it was the rule to give opium before operations.

September 3, 1828, M. Girardin read to the Academy of Medicine a letter addressed to his Majesty, Charles X., describing surgical anæsthesia by means of inhaled gases.

Richerand suggests drunkenness in reducing dislocations. Patients, while dead drunk, have been operated upon painlessly, and a dislocated hip was thus reduced after a bottle of Port wine. Haller, Deneux, and Blandin report like painless results in surgical and obstetric experience.

Baron Larrey, after the battle of Eylau, found in the wounded who suffered amputations a remarkable insensibility, owing to the intense cold. Of late years congelation has become a recognized agent of local anæsthesia.

Hypnotism is a very remarkable cerebral condition, by which surgery has been rendered painless. It is the grain of truth upon which the fallacies of mesmerism, animal magnetism, and the rank imposture of so-called spiritualism have been based.

The experiments of the Grotto del Cane are familiar, as also are recoveries from accidental asphyxia after complete insensibility.

About a year ago, two healthy men, at my request, inhaled atmospheric air from a common gas-bag. As carbonic acid replaced the oxygen, they both became livid, and, to every external sign, utterly insensible. One was really insensible; the other nearly so, but, being a plethoric subject, it was deemed prudent in his case not to press the inhalation further.

Nitrous oxide after a time asphyxiates, owing to the chemical combination of its gases—on that account parting reluctantly with its oxygen. But it also exhilarates, and its anæsthesia is probably something more than a condition of asphyxia.

These facts show that from time immemorial the world has been in possession of an anæsthesia which was occasionally resorted to, and not unfrequently amounted to complete insensibility. But, as a rule, the following propositions held good in respect to it.

1st. It could not be relied on to affect everybody.

2d. It was often insufficient.

3d It was sometimes dangerous.

What surgeons and patients needed was an inevitable, complete, and safe condition of insensibility; and this they were soon to have.

The moment arrived. In three months from October, 1846, ether anæsthesia had spread all over the civilized world. No single announcement ever created so great and general excitement in so short a time. Surgeons, sufferers, scientific men, everybody, united in simultaneous demonstration of heartfelt mutual congratulation.

It is to be regretted that no single individual stood out clearly, at this time, to receive the homage and gratitude of the world.

Nothing like the new anæsthesia had been known before. Whatever has been devised since has been a mere imitation and repetition of this—I may almost say, with no single substantial advantage over it. Our English friends, with a pardonable pride in matters of scientific discovery, not unfrequently formulate their convictions thus: "A. had indeed shown this, and B. that; but it was reserved for our own C. to make the imperishable discovery." It is probable that the average Englishman still believes that modern anæsthesia is identified with chloroform. But the discovery of a practicable, safe, and efficient means of insensibility had been made a year before chloroform was thought of, and nothing important has been added to it since. Chloroform is at a first inhalation of an agreeable odour, more portable and less inflammable than ether, qualities which eminently adapt it to army use; but it will do nothing that ether does not do as well, and is sometimes, though quite rarely, it is true, followed by fatal consequences.

We are now to examine the claims of the three individuals mentioned, to the discovery of the new anæsthesia. Let us look first at those of Dr. Wells. And as preliminary to the examination of his claim, let us here revert to an interesting point in this history. It is impossible to read the annexed statement, familiar though it be, without renewed amazement that this great blessing to animal existence was distinctly offered to scientific men, and as distinctly neglected by them for half a century.

The following are the words of Sir Humphry Davy, at the beginning of the present century; half the century had nearly elapsed before they were again thought of:—

"In one instance, when I had headache from indigestion, it was immediately removed by the effects of a large dose of gas" (nitrous oxyde), ... "though it afterwards returned, but with much less violence. In a second instance, a slighter degree of headache was wholly removed by two doses of gas.

"*The power of the immediate operation of the gas in removing intense physical pain I had a very good opportunity of ascertaining.*

"In cutting one of the unlucky teeth called *dentes sapientiæ*, I experienced an extensive inflammation of the gum, accompanied with great pain, which equally destroyed the power of repose and of consistent action. On the day when the inflammation was most troublesome, I breathed *three large doses* of nitrous oxyde. The pain always diminished after the first three or four inspirations; the thrilling came on as usual, and uneasiness was for a few minutes swallowed up in pleasure. As the former state of mind, however, returned, the state of organ returned with it; and I once imagined that the pain was more severe after the experiment than before. ......"

Towards the conclusion of his book he adds:—

"*As nitrous oxyde, in its extensive operations, appears capable of destroying physical pain, it may probably be used with advantage during surgical operations in which no great effusion of blood takes place.*"

The great discovery was here clearly pointed out to every tyro in medicine and chemistry. There weré three experiments, of a completely original character, and with a new agent, in a direction to which contemporaneous attention was not, as afterwards, leaning. Upon these an original hypothesis was methodically constructed and distinctly enunciated in print. It only remained for somebody to test this hypothesis, this guess, and to convert the guess into a certainty. But neither dentist nor surgeon responded, and the world for nearly fifty years attended exhibitions where people danced, laughed, and became unconscious, while hospital patients were undergoing amputation, alive to all its agony.

In 1844 Horace Wells appeared, exactly repeating the hypothesis that Davy had printed. Whether he got this idea from Davy, or from his friend Cooley, or from his own brain, is not to the purpose. Davy had announced it, and the scientific world knew it. Did Horace Wells convert into a certainty the probability of Davy? He did not. He signally failed to do so, and, mortified by his failure, he gave up all attempts in that direction. More than two years elapsed, and the ether discovery was made and completed. Then, and then only, was Wells stimulated to renewed effort. But it was too late; the discovery of a surgical anæsthetic had been made.

These facts deserve a careful examination. Wells had experimented in about "fifteen cases," and with varying success. This he states distinctly in his first reclamation, before his claim had expanded so as to embrace, as it afterwards did, the whole anæsthetic discovery.

He was "so much elated with the discovery" (to use his own words) "that he started immediately for Boston," and obtained from Dr. Warren permission, as Dr. Morton, imitating Wells, subsequently did with ether, to exhibit the anæsthetic properties of gas before the medical class. Dr. Warren was the principal New England surgeon of that day, and it was the obvious thing to do. This experiment, which was in tooth-pulling, was an utter failure, and was called, as Wells says, "a humbug affair." He was completely discouraged; went home, told his friends that the gas "would not operate as he had hoped," and wholly ceased to experiment, from the date of his failure in Boston, in December, 1844, until the spring of 1847, after he returned from Paris, an interval of more than two years. Wells's want of success can now be satisfactorily explained. He had, through Colton, in following Davy's instructions, made use of the traditional exhilarating-gas-bag, and of Davy's exhilarating dose. This volume of gas is inadequate to produce anæsthesia with any certainty; and Wells failed to suggest a larger dose. This small omission closed his chances. He narrowly missed a great invention. Inventors, by thousands, have missed inventions as narrowly.

Modern dental insensibility by nitrous oxide is unfailing, because the volume employed is much larger. It is also usual to exhale it into the atmosphere. Horace Wells had no hand in this method, of which the first demonstration was in a breast excision performed by myself at the Massachusetts General Hospital, in April, 1848, by means of about sixty gallons of gas, as a substitute for ether.

From all this it will be seen that Wells did not, as has been claimed for him, "discover that the inhalation of a gaseous substance would *always* render the body insensible to pain during *surgical operations*," but only that it would *occasionally* do so; and until long after the ether discovery, his experiments were *not "surgical operations," but only tooth-pulling*. Wells's anæsthesia had no value to patient, dentist, or surgeon. In endeavouring to trace dental anæsthesia, as Davy had directed, from toothache to tooth-pulling, his experiments unfortunately and erroneously showed that what availed in Davy's hands for toothache would not always avail for tooth-pulling. His slight, but fatal, error of using an inadequate volume of gas damaged the knowledge of his day; so that a scientific person who had read Davy's encouraging and unqualified prediction, based upon his three successful experiments, would have been discouraged and thrown off the track by witnessing Horace Wells's contradictory results.

Wells returned from Boston to Hartford. Having hoped much from anæsthesia as a money speculation, he now left it for more promising enterprises. He got up a panorama or exhibition of Natural History at the City Hotel at Hartford; initiated an extensive business in the sale of patent shower-baths; and somewhat largely invested in cheap copies of Louvre pictures painted in Paris, to be framed and sold by auction in this country. This carried him to Paris about December, 1846; an important event in his career, as we shall presently see.

Before he went, however, the ether discovery was made (October, 1846), and he received from Morton the following letter:—

"BOSTON, October 19, 1846.

"FRIEND WELLS. Dear Sir: I write to inform you that I have discovered a preparation, by inhaling which a person is thrown into sound sleep. The time required to produce sleep is only a few moments, and the time in which persons remain asleep can be regulated at pleasure. Whilst in this state the severest surgical or dental operations may be performed, the patient not experiencing the slightest pain. I have perfected it, and am now about sending out agents to dispose of the right to use it. I will dispose of a right to an individual to use it in his own practice alone, or for a town, county, or State. My object in writing to you is to know if you would like to visit New York, and the other cities, and dispose of rights upon shares. I have used the compound in more than one hundred and sixty cases in extracting teeth, and I have been invited to administer to patients in the Massachusetts General Hospital, and have succeeded in every case.

"The Professors, Warren and Hayward, have given me written certificates to this effect. I have administered it at the hospital in the presence of the students and physicians, the room for

Hartford Oct 20. 1846

Dr Morton Dear Sir Your letter dated yesterday is just received, and I hasten to answer it for fear you will adopt a method in disposing of your rights which will defeat your object. Before you make any arrangements whatever, I wish to see you. I think I will be in Boston the first of next week — probably Monday night. If the operation of administering the gas is not attended with too much trouble and will produce the effect you state, it will undoubtedly be a fortune to you provided it is rightly managed

Yours in haste
H. Wells

operations being as full as possible. For further particulars I will refer you to extracts from the daily journals of this city, which I forward to you.

"Respectfully yours,
Wm. T. G. Morton."

To this Wells returned the following remarkable and conclusive reply:—

"Dr. Morton. Dear Sir: Your letter, dated yesterday, is just received, and I hasten to answer it, for fear you will adopt a method in disposing of your rights which will defeat your object. Before you make any arrangements whatever, I wish to see you. I think I will be in Boston the first of next week, probably Monday night. *If the operation of administering the gas is not attended with too much trouble, and will produce the effect you state, it will undoubtedly be a fortune to you, provided it is rightly managed.*

"Yours, in haste,
H. Wells."

(*A fac-simile of this letter is here appended. The original is in the collection of the Massachusetts Historical Society.*)

Wells would not have thought that Morton's "operation" of "administering" his so-called "gas" (meaning ether) would prove "a fortune" to him, if his own results of two years before had, in the opinion of himself or anybody else, any considerable anæsthetic value.

The rest is briefly told. Wells soon after sailed for Europe, to prosecute the picture business already mentioned. The distinguished American dentist, Brewster, resident in Paris, hearing of his brother dentist Wells, sent to him, "begging him to call on him," and asked him, "Are you the true man?" "His answers, his manners," writes Brewster, "convinced me that he was." "*Dr. Wells's visit to Europe,*" writes Brewster again, "*had no connection with his discovery*, and it was only after I had seen the letters of Drs. Ellsworth and Marcy that I prevailed upon him to present his claim to the Academy of Sciences, the Academy of Medicine, and to the Parisian Medical Society." The quarrel of Jackson and Morton was Wells's opportunity, and Brewster's persuasion thus secured for him a European hearing.

Thoroughly uneasy, Wells returned home (March, 1847). The world was everywhere ringing with ether announcements. From this period his claim rapidly expanded, until it embraced the whole discovery, unsettled his business relations, embittered his life, unhinged his reason, and he at last died, in New York, a sudden death, after extraordinary acts which led even to his arrest, but for which he could not be considered responsible.

Thus perished Wells, volatile, ingenious, enterprising; an experimenter, like scores of others, in the field of anæsthesia, but, like them, unsuccessful in establishing anything of value. So far as his labours went, he left scientific knowledge, as well as its application to art, just about where Davy had left it half a century before. But he had kept the subject alive, and had unintentionally planted a seed in the mind of his ambitious partner which yielded fruit.

We now come to the claims of Dr. Jackson and Dr. Morton; and these, for convenience, we will consider together.

It is significant that Wells, Jackson, and Morton were all in contact at some period of their anæsthetic experiences, of which they shared in some degree a common knowledge. Wells, while in Boston, visited Jackson's laboratory; and Jackson says that he knew of Wells's experiments; and it should be observed that his advice to Channing and Peabody was after Wells's visit. Morton had been Wells's partner in dentistry, and boarded at Jackson's house In 1844 he had been a student of Jackson, who testifies in a certificate that he was a "skilful operator in dentistry," and had "studied the chemical properties of the ingredients required for the manufacture of artificial teeth."

Gas and ether were long familiarly known to produce effects so similar that whoever thought of one as an exhilarator or anæsthetic must have thought of the other. For example while in College, in 1837, I twice made a gasometer of nitrous oxide, and then substituted for it ether, as affording equal exhilaration. Brewster, in 1847, said, "It required neither a physician nor a surgeon to tell . . . . that ether would produce insensibility, ... as there is scarcely a school or community in our country where the boys and girls have not inhaled ether, to produce gayety, and many are the known cases where they became insensible." In short, gas and ether experiences were very similar. Wells had been, at the outset, distinctly advised to try ether, but elected gas as less dangerous. If, when afterwards "disheartened by the failure" of his gas experiment in Boston, he remembered ether, he doubtless thought it would be hardly worth his while to recur to an agent so similar in its effects.

In September, 1846, Jackson and Morton had their well-known interview. At this interview Jackson made a suggestion which was soon followed by a discovery, and by a controversy concerning the value of this suggestion. Jackson claimed that the suggestion was the whole discovery. Morton took the extreme opposite ground in behalf of his experiments. These he alone had conducted, and, while Jackson beyond all question kept aloof, he, recognizing their generally conceded danger, had gone on with them, notwithstanding, and proved what was before only suspected.

The interview was briefly this; and as it is the only point at which Jackson touches the progressive line of Morton's investigation, it should be stated strongly for Jackson. On the 30th of September, 1846, Dr. Morton went to the laboratory of Dr. Jackson, whom he knew well, having been a student under him, and recently in his house, and took from a closet an India-rubber gas-bag. In reply to an inquiry of that gentleman, he said, in substance (and all that Jackson claims Morton to have said may be admitted), "that he meant to fill the bag with air, and by its aid extract the teeth of a refractory patient." A conversation ensued upon the effects of the imagination, also, among other things, of nitrous oxide, in producing insensibility. Jackson treated Morton's proposition lightly. He told him to go to an apothecary and procure sulphuric ether—the stronger, the better—which would produce the insensibility he desired. The ether was to be spattered on a handkerchief and inhaled; in a moment or two perfect insensibility would be produced. "Sulphuric ether," said Morton, "what is that? Is it gas? Show it to me." Jackson showed him some ether, and after further conversation Morton went to procure it. Such was the substance of the interview at which Morton professed ignorance of ether, and Jackson entire knowledge of it. Jackson's knowledge and Morton's alleged ignorance we may now consider.

Dr. Jackson, if we may judge from his later attitude, was not indifferent to renown; nor was he regardless of the suffering of other people. He claims to have discovered, in 1842, that ether insensibility was infallible, thorough, and safe, and yet he cared so little for his reputation that he did not publish his discovery; and he forgot his humanity so far, that he allowed patients, the world over, to encounter the agonies of amputation during four years more. This extraordinary circumstance cannot strengthen belief in the fact of the discovery at this early date.

[1] It is to be observed here, that, if the patient was intelligent enough to obey instructions, and if Morton really meant to administer air, the patient would have become insensible by asphyxia. (See p. 79.)

In a long communication to Humboldt, in 1851, and in certain other communications to learned foreign societies and individuals, Jackson lays great stress upon the elaborate mental process which enabled him to construct, in 1842, an hypothesis of insensibility, based upon the distinct functions of nerves of motion and sensation, superficial and deep sensibility, etc. But the more studied and complete this hypothesis, and the more circumstantial the evidence of its careful elaboration, the more remarkable is it that it was laid on the shelf for four years. Without questioning the entire honesty of Dr. Jackson's convictions, it is safe to say that it is difficult to measure the original strength of any belief which has lain dormant for four years, especially if that belief has since proved to be a valuable truth. Dr. Jackson claims, indeed, to have mentioned his hypothesis to several persons during this interval; but this testimony, if carefully examined, relates chiefly to a narrative of his chlorine gas experiences. In fact, some of Dr. Jackson's immediate family had, during this same interval, in 1844, submitted to painful tooth extraction by Morton, and yet no anæsthesia was mentioned then. The hypothesis seems to have had for several years a precarious existence.

The only striking testimony is that of Peabody, a pupil, who was advised by Jackson to employ ether in having a tooth drawn, but who, after consulting his father, an accomplished amateur chemist, was deterred from doing so by the reputed danger, which Jackson's suggestion did not outweigh. But this advice was a year after Wells's experiments and failure in Boston, and his conversations at that time with Jackson. Whenever tooth-pulling was brought to Jackson's notice, how could he fail to think of Wells's experiments with gas? And who could think of gas as an inebriator, without its co-inebriator, ether? the obvious and only possible conclusion being, that what gas had done, ether might do, namely, sometimes succeed and sometimes fail. I have no doubt that this thought did occur to Jackson's mind when tooth-pulling happened to be talked of, especially after Wells's experiments.

It also occurred to Morton's mind. He knew more of Wells and of his varying experiments than Jackson did, and there is no question that Morton the dentist, oftener than Jackson the chemist, dwelt upon painless dentistry. His business was "mechanical dentistry," making sets of teeth; and he was daily suffering in purse because patients feared to have their teeth drawn. "I will have some way yet," said Morton to Gould, in August or September, 1846, a short time before the discovery, "by which I will perform my operations without pain." "I smiled," replied Dr. Gould, "and told him, if he could effect that, he would do more than human wisdom had yet done, or than I expected it would ever do."

Who that remembers the late Dr. Gould, cautious, accurate, truthful, judicial, the friend and brother scientist of Dr. Jackson, will doubt this sworn testimony? It was Dr. Gould, who, when his wife told him of rumors that Dr. Morton had drawn a tooth without pain, under the influence of something inhaled, exclaimed, "Yes, that can be done; ether will do it." So obvious was the transition from gas inebriation to ether inebriation, from gas insensibility to ether insensibility, in the mind of one who happened to have his attention drawn to the subject of anæsthesia by inhalation.

Morton knew of gas, and of Wells. He was in eager pursuit of anæsthesia. He believed in it. If he knew, also, of ether, he was, in all human probability, on the verge of discovery. Did he know of ether?

It is fortunately established beyond all question that Morton made long inquiries about ether in July, and also, a short time before the October experiments. If we reject the evidence of Wightman and Metcalf on this point, both of them disinterested, accurate, reliable, we must reject all human testimony.

Wightman the philosophical instrument-maker, afterwards Mayor of Boston, narrates a long conversation with Morton in the cars, when moving his family from his country residence to Boston. During this conversation he informed Mrs. Wightman, who asked him who Dr. Morton was, "that he was a dentist who was experimenting upon the relief of pain in dental operations." Wightman fixes the date, September 28th, of this journey, by items of expense charged in his day-book, of which the leaf was produced, as part of his sworn testimony. This enables him to fix the time of several previous conversations with Morton, concerning mesmerism, the effects of the imagination, etc., and especially of one as to whether rubber or oiled silk bags would hold "common ether."

Metcalf, the well-known apothecary, a scrupulously conscientious witness, equally substantiates the date of a conversation half an hour long, about ether, with Morton, who held in his hand a bottle of it which he had brought with him. In this conversation, "entirely about the inhaling of ether," interspersed with anecdotes of exhilaration and insensibility, Metcalf told Morton of "a person to whom he had given it, who was exceedingly wild, and who injured his head while under the influence of it, and did not know, when he got over the influence of the ether, that he had hurt himself, until it was called to his attention." "Morton," Metcalf testifies, "when he went away, knew as much about it as I did, for I gave him all the information which I had." Metcalf sailed for Europe in July, 1846, just after this conversation. While in Italy, in the early part of 1847, he took up a newspaper announcing the discovery, by a Boston dentist, of insensibility through ether inhalation. "I said at once," testifies Metcalf, "that I was sure Morton must be the man, for he was engaged upon ether before I left home; and that I now knew why he had been so curious, and at the same time shy, in his conversation with me." To those who know Mr. Metcalf this evidence has all the weight of personal experience.

There can be no question that Morton knew about ether. How much he knew about it is less important. But it should be mentioned that he claims to have made repeated experiments with it upon animals in the summer of 1846.

Morton's explanation of his professed ignorance of sulphuric ether was this. During the summer, while boarding at Dr. Jackson's, he had heard frequent and protracted expositions of Jackson's claim to the invention of the electric telegraph, then recent, and the important features of which Jackson was satisfied he had communicated to Morse during an ocean passage. Dr. Jackson had, indeed, a well-stored and suggestive mind, which made it more than likely that he had furnished information, of which Morse, while originating and mentally evolving a system of electric telegraphy, may have been glad to avail himself. A sharp public discussion, with pamphlets, ended with a verdict in Morse's favour. In going to Jackson, Morton feared that if there were any deliberate conference, Jackson might set up a similar claim of participation in his own search for painless dentistry, and therefore took the shortest way to exhaust his knowledge for his own benefit, without discussion. There can be little doubt that Morton was in this matter reticent, as Metcalf states, and intended to keep it a secret from his brother dentists. I am also inclined to believe that Morton at first cared little about the abstract question of discovery, and would have willingly left a large share of any honour unquestioned in the hands of Jackson. But when Jackson made a claim upon the patent, and the profits, beyond the amount to which Morton conceived him to be entitled, then he defined his own claim to the discovery. It may be stated in this connection, that the surgeons of the Massachusetts General Hospital, who had no interest whatever in this difference, and could have none, were friends of Jackson, and strangers to Morton. They yielded, when it became necessary to take sides, only to their deliberate conviction of the justice of Morton's claim.

It will be advantageous, at this point, to take a general view of the "ether controversy." For this purpose, I find I can do no better than to quote the following letter, written, when the occurrences were fresh, to the Hon. Robert C. Winthrop, in Washington:—

January 26, 1848.

DEAR SIR: I believe most fully, that Dr. Morton deserves any reward Congress may grant to the discoverer; because, although many people have *thought* that a man could be intoxicated beyond the reach of pain, Dr. Morton alone *proved* this *previous possibility* to be *a certainty*, and *safe*. A diagram will make the matter plainer than words:—

Before October, 1846, who made the suggestion?	Discovery in Oct. 1846. Consecutive experiments by Morton.	After October, 1846, *Morton alone* took the responsibility of danger, and proved that it was
Here is the only ground of dispute.		1st. *Certain.*
		2d. *Safe.*

The two last points, namely, the consecutive experiments, and their confirmation, *which nobody disputes to Morton*, make him, in my eyes, the discoverer. The only doubt is, who made the *suggestion?* *To me this is of no importance.* Dr. Jackson says, "I did. I told Mr. Morton to try the experiment; and unless I had so told him, he would never have tried it." Dr. Jackson adds: "I first tried ether when I was suffering from chlorine, in 1842. I afterwards recommended it to Mr. Peabody." But Dr. Morton confutes even these positions. He says to Dr. Jackson: "1st. I show, by the evidence of Dr. Gould, Mr. Wightman, and Mr. Metcalf, that I was experimenting with ether before the interview in which you claim to have brought it to my notice. 2d. In 1842 you only re-discovered what was before clearly in print in Pereira's *Materia Medica*. 3d. You claim that you told Mr. Peabody what you *knew* of ether. Now you could not *know* it. You have stated all your grounds of deduction, and the widest inference you could draw from them is, a *suspicion* of the properties of ether; and a *suspicion* in science, an *unconfirmed theory*, amounts to nothing. Finally, what you claim to have discovered in 1842 you kept to yourself during four years. Do you expect the world to believe you knew its value? Do you expect it to reward you for letting people suffer during that length of time? Besides, the suggestion of anæsthetic agencies occurred to Davy; especially was it followed out, though unsuccessfully, by Horace Wells, who, disgusted with failure, abandoned his attempts." These and others had hypotheses as well as Dr. Jackson. Morton alone proved the hypothesis. Without Morton there is no evidence that the world would have known ether to the present day. I believe this covers the ground of important argument and difference in the pamphlets. . . . .

Respectfully your ob't servant,
HENRY J. BIGELOW.

At the interview referred to, Jackson's partisans tenaciously insist that he assumed direction, as a physician might have done, of the administration of a remedy; while Morton acted only as a "nurse." Let us examine, then, the often-quoted "nurse" argument with which the opponents of Morton have endeavoured to handicap him at the outset. Here is its fallacy. A physician, by common understanding, possesses a positive knowledge of the effect of his remedies, in advance of their administration. Such knowledge was impossible to Jackson. Again, a physician is employed expressly to direct, and a nurse to obey. Under these circumstances a nurse is not likely to get much credit for originality, which is in fact absolutely excluded by the terms of her contract. There was no such contract here. Morton was not a mere agent, without preconceived plan, automatic in action, and irresponsible as to results. On the contrary, he was already conducting an independent investigation. He was in pursuit of anæsthesia when he went to the laboratory of Jackson, with whom the subject, even admitting his claim, had slumbered for years. He had been before, and in the same way, to books, apothecaries, instrument-makers, in short, to various usual and available sources of knowledge, as is customary with every investigator or discoverer. The purpose, the investigation, the patient, the discretion, the responsibility, were all his. Morton was not a "nurse."

Morton had a combination of qualities such as few other men in the community possessed. Fertile in expedients and singularly prompt in execution, he was earnest and persevering beyond conception. His determined persistence is remembered even at this interval of time, as having been a terror to his best friends. Nobody denies that Morton, recklessly and alone, faced the then supposed danger attending ether stupor. If all accredited scientific opinion had not been at fault, and in the case of any fatal result, he would have infallibly been convicted of manslaughter, with little probability that anybody would have come forward to say, "The responsibility is not his, but mine."

In fact, Dr. Jackson endeavoured at this time, by word and deed, to keep both himself and his reputation clear of Morton, as a reckless and dangerous experimenter. The only operations under ether witnessed by him during the first three months were on November 21, 1846, and January 2, 1847; and a part of this time he was absent from the State.

"But," it has been a hundred times said, "Jackson made a suggestion, and Morton used it." It is evident that we cannot escape some discussion of the relation of "suggestion" to "discovery," no matter how little the suggestion may be intrinsically worth, or how fortuitous its success. The simple question is, What was the actual value of Jackson's suggestion to Morton at the time it was made—I distinctly mean before Morton had handled it?

A suggestion in science varies in value as much as a suggestion in common life, where advice is not always sound advice. It may, indeed, turn out fortunately, like a suggestion, for example, to wager money on the ace of spades. But because it may prove so, the advice does not necessarily imply merit in the adviser. This is an extreme case. At the other extreme is a mathematical certainty, such, for example, as that twice two make four: a truth the value of which is not afterwards augmented for intelligent people by a material test or demonstration that twice two apples make four apples. A similar example of mathematical certainty is that calculated by Leverrier (liable only to the instrumental fallibility of dividers, telescopes, and equations), which did not become more absolute, nor more true, when Galle saw the planet where Leverrier told him it must be. Mathematics are unerring, while predictions in physiology are as uncertain as predictions about the weather a week hence. Yet it has been argued that Jackson was Leverrier, and Morton Galle.

Jackson's alleged hypothesis, before Morton took hold of it, had only the value of a lottery ticket, which, through Morton's unaided, dangerous, and acknowledged efforts, drew an immense prize—or of "the cast of a die"—"for in that light," says Watt, whose name is identified with the history of steam, and the soundness of whose practical views no one will dispute, "I look upon every project that has not received the sanction of repeated success." A statement of the grounds upon which this view is based will enable others to draw their own conclusions. Let us begin with Jackson's experiments.

"Having, in 1841–2," says Jackson, "got my lungs full of chlorine gas, which nearly suffocated me, I immediately had ether and ammonia brought to me, and alternately inhaled them, with great relief." The next day, still suffering, and "perceiving a distinct flavour of chlorine in my breath," the experiment was repeated "with perfectly pure washed sulphuric ether," and "with entire loss of feeling." "All pain ceased." He "fell into a dreamy state, and became unconscious of all surrounding things."

In other words, he inhaled ether, until, seated as he was "in a rocking-chair, with his feet in another chair," he fell asleep. So far there was nothing new. Pereira, for example, says (*Elements of Materia Medica*, etc., London, 1839, pp. 210-211):—

"The vapour of ether is inhaled in spasmodic asthma, chronic catarrh, and dyspnœa, whooping-cough, and *to relieve the effects caused by the accidental inhalation of chlorine gas.*" "When *the vapour of ether sufficiently diluted with atmospheric air* is inhaled, it causes irritation about the epiglottis, a sensation of fulness in the head, and a succession of effects *analogous to those caused by the protoxide of nitrogen* (vide p. 156), *and persons peculiarly susceptible to the action of the one are also powerfully affected by the other.* (*Journ. Science*, vol. iv. p. 158.) *If the air be too strongly impregnated with ether, stupefaction ensues.*"

Such was contemporaneous knowledge. Jackson's experience was identical with that recorded by Pereira. And Pereira further distinctly calls attention to the similar effects of inhaled gas and inhaled ether, and to the stupefaction caused by ether.

Pereira adds, in regard to the danger of this condition:—

"In one case this state continued, with occasional periods of intermission, for more than thirty hours; for many days the pulse was so much lowered that considerable fears were entertained for the safety of the patient. (*Op. cit.*) In another case, an apoplectic condition, which continued for some hours, was produced."

I shall revert to this subject of danger.

But Dr. Jackson alleges that he now advanced a step

further, and logically inferred the entire safety and inevitable certainty of ether anæsthesia, in all cases, and during the severest surgery. This was an unwarrantably wide inference, as we shall see.

Further still, Dr. Jackson claims to have invented a method upon which the success of his hypothesis largely, as he supposed, depended; and he offers this method as evidence that he made the hypothesis and the discovery. It is plain that a discovery may be in this way rendered more probable. Whoever is in possession of a new method is more likely to find the way to a new truth. But if the alleged method proves to be partly erroneous and partly an old and familiar matter, then the proof of alleged discovery is no stronger because of it. In fact, just so far as the hypothesis was apparently stronger by reason of the method, it becomes weaker when the method falls to pieces. Let us, then, examine this alleged method.

Dr. Jackson has from the first insisted upon two points, as peculiar to his invention, and essential to his discovery. By these safety and certainty are secured, and a neglect of them explains previous uncertainty and danger. They are,

1st. Purity of the ether;

2d. An admixture with it of atmospheric air.

If these conditions are either not essential or not new, we find ourselves, at least so far as method is concerned, where Pereira left us.

From time immemorial, the familiar way of inhaling ether has been with air from a handkerchief. Pereira, as before quoted, distinctly stipulates for air. Says another writer, "Animals breathe oxygen. Without oxygen a man must die. Ether would not have saved Desdemona." This needs no ghost for its enunciation. But the fact is, in etherizing by a sponge or cloth, the difficulty is as often to exclude air enough as to admit air enough. Some good authorities maintain that there is advantage in its exclusion—that the insensibility from gas, for instance, is due to asphyxia. Even in etherizing, they aim to take advantage of this condition by restricting the air supply. The French ether-bag, still in use, is expressly arranged for this purpose. In short, while it requires especial effort to exclude air, partial asphyxia is not dangerous; and a claim to the discovery of the safety or certainty of ether stupor based upon the admission of air cannot stand.

A claim based upon pure ether is equally void. Tolerably pure ether is better, but by no means essential, for safe insensibility. The specific gravity of pure ether is not very far from 0.718, and is difficult to attain; that of our usual inhaling ether of Powers & Weightman, 0.724; and of Squibb (*fortior*) about the same. The ether of the old Pharmacopœia, and of the shops in the year 1840, was not far from 0.750; and this is a very practicable anæsthetic. Its slight adulteration with alcohol and acid is not especially deleterious, when inhaled. In fact, anæsthesia resembles dead-drunkenness, which is equally possible with brandy, or with brandy and water. The real danger was not from impure ether, but from over-inebriation—a danger which exists to-day with the best ether.

The discovery was not in the admission of air, nor in the use of a particular quality of ether. It was, that the inhalation of ether, which had been familiarly resorted to for years, could be prolonged beyond the usual stage of exhilaration to a stage of stupor, possessing the *complete* insensibility of a dangerous coma, yet, unlike that condition, *safe;* and that all this could be effected *in every case.*

The history of ether anæsthesia was the gradual discovery of the following facts:—

1st. That ether inhaled produced, capriciously, in certain instances, unconsciousness. (Old.)

2d. That it possessed the power of producing stupor in every case. (New.)

3d. That this stupor could be exactly graduated as to time. (New.)

4th. That it could be increased or diminished, and arrested short of danger, and so made safe. (New.)

5th. That there were certain infallible indications of danger. (New.)

6th. That, while thus controllable and safe, it could be made at will so profound that even amputations should not be felt. (New.)

All that is new belongs to Morton.

A person inhaling ether from his handkerchief would drop it when his hand and senses were benumbed. As air entered the lungs he would revive. Such was Jackson's self-experiment, already covered by Pereira. But a second person, who should now seize the handkerchief and compel the continued inhalation to the stage of stupor, might obviously make a valuable study of this new ground. He could draw his patient's tooth, or amputate his leg, and thus measure the insensibility. He could repeat the experiment until satisfied that insensibility could be always attained, and that it was safe. All these experiments Morton tried upon others, and when he had tried them the discovery was made.

Jackson virtually claims that his inference extended to the *universal efficacy*, the *completeness*, and the *safety* of the stupor.

Of the *universal efficacy* of the new stupor Jackson could have no valuable opinion. Such knowledge was possible only through observation of many cases, with an experience which could say, "Administer ether as for exhilaration. Protract its inhalation beyond this usual stage, and you will inevitably reach an ulterior condition of stupefaction. During this process a patient may struggle with a giant's force; or perhaps will tell you of his extreme distress; or after five or ten minutes of inhalation will satisfy medical bystanders, as patients sometimes do now, that he is not amenable to ether. Another will assure you, with apparent reason, that he is dying; or, livid with asphyxia, will compress his lips and cease to breathe; or may, for a long time, alternate between lividity with a failing pulse, and arterialization with a partial return of consciousness.[1] But, through all these apparently alarming indications, urge the process discreetly, and you will ultimately and inevitably reach a stage of stupefaction."

Dr. Jackson could not say all this, because he knew nothing of it; nor did anybody else, until Morton established the fact that there was no exception to the potency of ether. These and other contingencies, once startling, still occur with the best ether, and the experience of a quarter of a century. Had Morton been a timid or a discreet man, anæsthesia might have been delayed beyond the present generation. Morton compelled inhalation in spite of indications to arrest it, incurred the responsibility of doing so, and is entitled to the credit.

The completeness of the insensibility—its adequacy, for example, in amputations—was settled only by repeated observation, with varying results, but gradually accumulating evidence: from the tooth first painlessly extracted, through several failures to comparative success in dentistry; then the removal of a venous tumour of the jaw, where the patient was doubtfully insensible; then a fatty tumour of the arm, with complete insensibility; and finally, after an inconsiderable operation or two, the amputation of a leg, practically successful. Everybody awaited this final test by amputation, and then only was the accumulated evidence

[1] Such patients are familiarly known to our hospital attendants as "bad etherizers." Some of them are inebriates. If they return for a second operation, they seldom fail to exhibit the same symptoms as before.

deemed conclusive. It was, indeed, beginning to be felt that the process was a safe one, and that it promised satisfactory results; otherwise the hospital surgeons would not have permitted this test of it in a patient weakened by disease. But this experiment, like the rest, so far as anæsthesia was concerned, was Morton's, and a part of his well-organized enterprise to investigate and establish the new insensibility. As this amputation has been repeatedly published as mine, it should be stated that it was performed by the late Dr. Hayward. That anæsthesia was employed at all on that occasion Alice Mohan has to thank me; although, if anæsthesia had not been employed in this particular instance, the test of amputation would have been delayed only till another should occur. The circumstances may illustrate the sort of obstacles Morton encountered. The looked-for opportunity had arrived; but, through various antagonistic influences, Morton, in spite of his earnest request, had been notified that ether would not be administered, and that he need not attend the operation. Of this he informed me the night before, and by arrangement with him I carried him to the hospital the next day, just before the operation, there to await events. A dose of laudanum had been administered, and Alice Mohan was carried to the amphitheatre, for operation without ether. I there strongly urged the employment of the yet ostensibly secret agent; partly on the ground that it then was really known, but especially from the consideration that humanity ought to supersede any doubts connected with professional etiquette. This and other considerations prevailed, and, after a delay of half an hour, Morton, whose presence had been till then unknown except by me, was brought up, and the patient was etherized.

The evidence of all this slowly accumulating anæsthetic surgery, at which Dr Jackson was not present, was claimed for him in virtue of the "nurse" argument. What light could a repetition of the chlorine experiment of Pereira throw upon the question whether a patient could sleep during an amputation?

Ether exhilaration was familiar; but, on the other hand, it was well understood that ether stupefaction was in certain cases dangerous. Physiologists had also found that the smaller animals very frequently died of it. Brodie wrote of the fresh discovery: "I have heard of this before, and had tried it on guinea-pigs, whom it first set asleep and then killed. The great question is, Is it *safe?*" This was, indeed, the question. Could danger be avoided? what was its exact character? and what were its indications?

On the second of January, nearly three months after the discovery, Jackson came to the hospital for the first time. To that date, he had been present at one operation only, that at the Bromfield House, November 21. He was not cognizant of current experiment, and brought voluntarily a bag of oxygen, which he urged upon the hospital surgeons as a necessary precaution against danger, erroneously supposed by him, at that late date, to be asphyxia, instead of over-inebriation, of which the essential indication is the pulse. It was some weeks after the discovery, that this source of danger, and its sign, were understood; and in the mean time Morton might have killed anybody. A patient was, in fact, in great danger from over-inebriation at the first private operation. He was inhaling, in the continuous way that was at first supposed to be essential to protracted insensibility, through a glass globe of ether, and long after insensibility was manifested. The operation was far from completed, when a bystander happened to feel the pulse. There was no special reason for doubt, inasmuch as the patient was, in general appearance, like all former thoroughly etherized patients. The pulse proved to be barely perceptible, and the patient to be etherized almost beyond recovery. The bystander, after repeated observation of other cases, published the fact, then first observed, that in ether anæsthesia the pulse stood as a beacon between safety and danger, between harmless inebriation and fatal narcotism. This was the discovery that ether was not dangerous; because this showed that its danger gives warning, and is under control. The operator was Dr. Dix, the bystander myself, and the discoverer Morton. To his impetuous, unremitting, reckless experimentation to establish anæsthesia, surgeon, bystander, patient, ether, and apparatus, were all for the time, and in that relation, subordinated. Morton had asked me to be present, because I was more familiar with the new process than anybody except himself, and for the purpose of aiding him, in emergency, with professional advice. But the anæsthesia was his. I assumed no responsibility. Had the patient died in a "stupor," as he might well have done, Morton was liable; and as the patient did not die, his was the credit. This was real danger. But there was other danger, more startling, though only apparent; such as prostration, "trance," or "mania," lasting for hours, and for which Morton was in one instance threatened with prosecution. What was Dr Jackson's responsibility? None whatever. He was then absent from the State. What had Pereira's chlorine experiment taught him about all this danger? Absolutely nothing; nor could it do so. And he could impart nothing.

In view of these facts, which leave Jackson standing upon the naked experiment of Pereira, we may fairly pause, and ask, What, in the fullest sense, were the exact significance and value of the suggestion made by Jackson to Morton?

Caleb Eddy says, in his sworn testimony, "I said to Dr. Jackson, 'Dr. Jackson, did you know, at such time, that, after a person had inhaled the ether and was asleep, his flesh could be cut with a knife without his experiencing any pain?' He replied, 'No, nor Morton either: he is a reckless man for using it as he has.'"

Waiving this testimony, it is clear that the whole of the

peculiar knowledge embraced in the suggestion of Jackson to Morton would have been accurately and fully conveyed in the two words, "Try ether." This suggestion should be fully credited to Jackson. Morton never questioned the fact of the suggestion; on the contrary, he at once proceeded to square the account. Jackson at first distinctly agreed to receive five hundred dollars, as full compensation for the assistance he had rendered. As the evidence grew, and the greatness of the discovery became more apparent, Jackson raised his demand to ten per cent. on the sales of patent rights. Later, when it was clear that the lottery ticket had drawn an immense prize, Jackson again increased his demand to twenty-five per cent., which Morton refused. The controversy was then opened.

I very early urged the inexpediency of a patent, if only on the ground that, like Whitney's cotton gin, for example, this invention was so valuable that the world would combine, as the event has shown, to take possession of it; and that the question of equivalent might safely be left to public generosity, which has generally recognized such debts. Ether anæsthesia was at first opposed on the ground of its danger, of quackery, of religion, and of professional etiquette. Much of the early opposition to a patent[1] came from dentists, who desired to use the new method without pay; and

[1] The first statement of the fact of operations under insensibility was a paper in the *Boston Medical and Surgical Journal*, Nov. 18, 1846, entitled "Insensibility during Surgical Operations, produced by Inhalation. Read before the Boston Society for Medical Improvement, Nov. 9, 1846, an Abstract having been previously read before the American Academy of Arts and Sciences, Nov. 3, 1846, by Henry Jacob Bigelow, M.D., one of the Surgeons of the Massachusetts General Hospital." A copy of this was sent by a gentleman to his friend, Dr. Boot, of London, and by him transmitted to several London surgeons. Their replies to Dr. Boot I have in my possession. This paper also announced the patent, and the connection with it of the names of Morton and Jackson.

they confused with it the question of humanity to suffering. But in those days dentists had secret methods to which they attached a money value, and which went far to justify both secrecy and patent. The question of secrecy should, however, be detached from that of equivalent. Dentists and physicians, lawyers and clergymen, dealers in food, heat, light, labour-saving methods, in short, in comfort, knowledge, and value, rightly exact a pecuniary equivalent, proportioned to their services, from those who can pay it without inconvenience. The more distinctly this is recognized, the better we shall understand the nature of real humanity, and the more readily lend assistance and charity to those who need them. Jackson was right in expecting a money return for the service he had rendered. The only question here is, How much he himself, at first, considered a fair equivalent for this service. This has been shown.

After it became evident that the patent was worthless, Jackson repudiated the division, and claimed the whole discovery, in virtue of the "nurse" argument. Under these circumstances Morton properly insisted that the "suggestion" could have been picked up from almost any source, by any man actively searching for painless dentistry, who knew everything about Wells's experiments with gas, and who also knew the familiar and similar action of gas and ether. Morton was right.

But Morton also urged that this was a discovery not in science, but in art; that surgical anæsthesia was due, not to any great scientific novelty in the long recognized ether insensibility, but to his having worked out the application of this insensibility to use in art, with enterprise and perseverance, through many details, in the midst of danger, till he gave to the world a perfected system of efficient and safe anæsthesia. Morton was again right.

When a discovery is great, not from the intellect invested in it, but because it ameliorates the condition of mankind, then its recognition has more of a business character, and the gratitude and honour bestowed by the world are more nearly an equivalent for value received. The world does not concede the equivalent until it has received the value; and it is apt to examine with business exactness the claims of those who profess to have acted as agents in the matter.

This suspicion of discoverers who do not appear until after the world has been made to recognize the discovery is justified by the fact that hardly an invention of importance was ever made known that it was not at once claimed, often simultaneously, from a variety of sources. This is not remarkable. The world, whether in science or in art, is built up to a certain point by the easy and wide transmission of knowledge. Upon this elevation stands a multitude of philosophers, engaged often in identical researches, and possessed of much information upon the subject in question. Each of these may honestly believe that his imperfect knowledge is the perfected knowledge in question. In such a case the world is liable for a short time to confound claims, to confuse the incomplete result of a few data with the completed demonstration from many, the unproved with the indisputable, theory with fact. Recognize this fallacy, and the question of invention is comparatively simple. Yet it is not recognized. There is at this day the same claim to priority in invention as existed half a century ago. The writer of a *Life of Fulton* then said: "Those who question Mr. Fulton's claim are precisely those who have been unsuccessful in their own attempts; and it would seem that exactly in proportion as their efforts were abortive, and as they had thrown away money in fruitless experiments, their claims rose in their own estimation and that of their partisans." The witness—I believe before the House Commons—probably did not overstate the matter when he gave it as his opinion, that, if a man were to show that he had found a road to the moon, his neighbours would testify, that, if they had not been there themselves, they knew several persons who were familiar with the road in question. It is hardly too much to say, that, at the present day, every invention or discovery having a large supposed value is systematically contested.

Morton, in his attitude of investigator, had a right to receive a hint or suggestion of greater solidity than that of Jackson, without impairing his merit as inventor. Every invention is preceded by such hints or suggestions derived from experiment, books, or people. Curtis (on Patents) says: "It is clear that many suggestions may have been made, or many hints taken from others, without invalidating the claim of a party to be considered as the author of the invention."

Of a hundred instances easily cited to illustrate this, let us confine ourselves to a medical one, that of the invention of vaccination to avert smallpox.

The young countrywoman of Sodbury said of smallpox: "I cannot take that disease, for I have had' cowpox." The Duchess of Cleveland, when Lady Mary Davis and other companions taunted her as likely to deplore the loss of that beauty which was her boast, as the smallpox was then raging in London, said that she had no fear about her beauty, for she had had a disorder which would prevent her from ever catching the smallpox. Were these discoverers? Surely, yes, if Dr. Jackson was one. In fact, the hint that they and others gave to Jenner in the vale of Gloucestershire, where he resided, embodied more knowledge of vaccination than Dr. Jackson's suggestions did of ether anæsthesia.[1] But

---

[1] The experiences of the milkmaid and the Duchess might easily have apprised them of their own immunity from smallpox. But Jackson, through his chlorine experiments, could have no evidence that ether stupor was capable of affecting all persons, or that it gave immunity from real surgical pain, or was free from danger. No self-experiment could touch these points.

neither the milkmaid nor the Duchess of Cleveland was ever honoured as a discoverer of truth, or an inventor of method, while Jenner was so honoured. The reason is obvious. They, like Dr. Jackson, asserted a doubtful fact, and had neither time nor inclination to pursue the subject; but Jenner, by multiplying instances of the cowpox inoculated like smallpox, which was already supposed to be, like smallpox inoculation, protective, conclusively proved that it was thus protective, and also safe. He did with cowpox what Morton did with ether.

The parallel in this case is very close. Jenner and Morton were in pursuit of what the scientific world regarded as a chimera. Because they believed in its possibility they encountered prejudice and opposition. They both received from others a hint, suggestion, or surmise, which afterwards proved to be capable of development into a truth of great value. This suggestion was based on narrow induction, and had, therefore, obtained no previous general credence. It had also slumbered for years at that stage of its development. The world had not believed it.

Jenner and Morton, both men of singular persistence and perseverance, took hold of this idea, of which, from the nature of their daily pursuits, they felt the immense value. But for them it might have slumbered indefinitely. They dragged it through till the world recognized it and them. This they did at the risk of injuring people's health, of killing them, and of being held responsible for so doing. Nobody has ever doubted that Jenner was the inventor of vacciuation, and nobody should doubt that Morton was the inventor of modern anæsthesia. Here the parallel ceases. The English people voted Jenner a reward of $150,000.[1]

---

[1] In decisions relating to discovery, unanimity is not to be expected. There can be none where partisanship and large interests are involved. The question, then, is, Where the weight of evidence

The world demands convincing demonstration—and not by an individual for himself or to himself, but to them and for them, with overwhelming clearness. Then, and not till then, it responds with acknowledgment, concession, or gratitude.

Sydney Smith, in the *Edinburgh Review*, insists on this. In fact, he wittily overstates the claim of the mere publisher of a novelty, when he says that "he is not the inventor who first *says* the thing, but he who says it so long, loud, and clearly, that he compels mankind to hear him."[1]

lies. Even Jenner, with no rival, encountered great hostility both to himself and his discovery. The House of Commons (June 2, 1802) voted him ten thousand pounds by a vote of 59 to 56, a majority of three only. A further sum of twenty thousand pounds was voted (June 29, 1807) by a vote of 60 to 47, a majority of thirteen. Morton's award of $100,000, for his patent, passed the Congressional Committee. It was arrested, not wholly by Jackson, but by the partisans of Horace Wells, who published what afterwards filled an octavo volume, containing little argument, but full of bitter invective against Morton, and promised, if only delay were granted, to make a conclusive case for Wells.

[1] It is easy to understand what is meant here. For instance: I have amputated more legs after applying a tight bandage from foot to hip before tightening the tourniquet, than in any other way. Similarly to the arm for a needle in the hand. A hundred surgeons have done the same thing. Sir Charles Bell went a step further, and announced the dry method in print. "I may here observe," he says, "that by the management of the tourniquet, blood may be lost or gained. If the limb be uniformly rolled before amputation, the veins are emptied into the general system, and blood is saved instead of being withdrawn. In a very exhausted state of the patient, it may be of service to attend to this." (*Great Operations of Surgery*, London, 1821, p. 58.) But Esmarch was so impressed with its importance, that he erected it into a system, and urged it upon the attention of every surgeon in the civilized world. To many the idea was new. In fact, to a considerable part of the surgical world Esmarch was the discoverer of an important truth. He deserves to be avowed as such, in requital of his pains to perfect

Thus put, however, the statement throws light on Jackson's prominence after the discovery. An unproved hypothesis in physiology, which was his whole claim, would usually be considered of little account. But Jackson knew the machinery of fame. As Wells by accident, so Jackson by his scientific relations, and through his friend and former teacher, Elie de Beaumont, got at once possession of the scientific rostrum of the French Academy of Sciences and other foreign learned societies. He also subsequently addressed Humboldt. He thus compelled the European world, even so far as Turkey, to listen to his exclusive statement that the whole was his. In the mean time, however, there was in Boston a scientific jury of the vicinage. When Morton's statement at last crossed the water, the

and to publish to the surgical world, a useful point, new to many of them. He is so recognized.

Antiseptic precautions in surgery are not new; but Lister published his views, as did Esmarch. Whether germs are essential to the theory or not, there can be little doubt that it is well to free a wound from coagula, and to wash it out with diluted carbolic acid or its equivalent. Then the subsequent free use of the antiseptic, by hindering decomposition without, tends to maintain vitality within the wound. These views have been enforced and brought home to the world by Lister, with the pertinacity of Jenner or of Morton. The world at large owes its attention to these points to Lister's announcements, and properly attaches his name to the antiseptic method.

Mere publicity is notoriety; but with merit it is fame; with originality it is discovery. To all these publicity is essential. He who keeps his discovery comparatively to himself discovers, or uncovers, nothing. If he claims to have done so, the world will scrutinize and suspect his claim.

The act of publication, indeed, adds little to the claim of Morton; his impregnable position being that he elaborated and perfected a new art. But facts like the above go far to show that the failure of Jackson for four years to publish his alleged guess, of itself extinguishes any claim to its recognition during that time.

French Academy did what it could at that late day, and awarded to Morton the same honour and recognition it conferred upon Jackson. It could do no less. It could then do no more. But had the case been at the outset reversed, and had Morton made a suggestion to Jackson, does anybody doubt that the humbler Morton, and his suggestion, would have been by scientific precedent wholly absorbed and assimilated?

All honour, then, to the inventor of the art of anæsthesia!— for there was little science in it. He found the practice of ether inhalation an amusement of chemical lecture-rooms and schools; he left it the sovereign anodyne of the human race in its moments and hours of agony. He found ether stupor as hazardously uncertain as was the narcotism produced by pouring down the opium "*à boire*" of Canappe; he left it as manageable and safe as the sleep that follows a dose of laudanum.

There is hardly an inhabitant of the civilized world but can remember some one of those nearest to him in whose experience the anguish of the knife or of disease, of birth or of death, has been assuaged by anæsthesia, perhaps converted into a pleasant dream. Yet he is willing to take this priceless boon as a gratuity from those whose sole patrimony it was, and who have been brought nigh to want that he might enjoy it. If the world should cancel a fraction of its debt, the family of the inventor could afford to be generous to the families of his former friends, who, without impairing his title to the discovery, contributed to his success.

Wells's sad story has been told. Morton fell with apoplexy, induced by a publication in behalf of Jackson, of a nature to prejudice a subscription then arranged in New York for his benefit. Jackson, it is to be feared, is at the present time hopelessly bereft of reason.

SURGERY.

BY

S. D. GROSS, M.D., LL.D., D.C.L. Oxon.,
PROFESSOR OF SURGERY IN THE JEFFERSON MEDICAL COLLEGE OF PHILADELPHIA.

10*

SURGERY.[1]

"According to this time it shall be said, what has God wrought?"

A CENTURY has elapsed since the American colonies, through their representatives in Congress, assembled in this city, absolved their allegiance from the British crown, and, after a struggle of seven years, attended with great sacrifice of blood and treasure, achieved their independence as a free and sovereign people. What the country has done since that eventful epoch, in the various pursuits and occupations of life, is a legitimate object of inquiry, especially at a moment when the Nation, now composed of forty millions of human beings, is about to celebrate its Centennial Anniversary, in a form intended to display, in the amplest manner, its mental and physical wealth. It is well that every profession once in a century should open its ledgers and examine its accounts to see how it stands with itself and with the world at large.

The progress of the arts and sciences is intimately associated with the intellectual development of the human race. No nation can be truly great, if unmindful of the sanitary condition of its citizens. Civilization and the arts of domestic life march hand in hand; as is the one, so necessarily

[1] The author tenders his acknowledgments to Professor Greene, of Portland, Professor Bell, of Louisville, Professor Johnston, of Baltimore, Dr. George A. Otis, and Dr. Harvey G. Brown, U. S. A., and Dr. Laurence Turnbull, Dr. William Thomson, and Dr. I. Minis Hays, of Philadelphia, for material embodied in this paper.

must be the other, so indissolubly are they interwoven and bound together. The refined and cultured physician has been an object of the deepest interest in every enlightened age and country. Even the American savage, who cares little for his physical comforts in his native wilds, has his "medicine man" to ease him of his pains when overtaken by disease or accident. In ancient Greece and Rome, those great centres of civilization in the Old World, the physician was held in the highest esteem and veneration. The beautiful and complimentary remarks of Cicero are familiar to every classical scholar, and meet with a ready response in the heart of every right-thinking person. The praises of the surgeon have been sung in poetry and heralded in prose the world over, and there has been no important military enterprise since the first great battle was fought in freedom's cause, in which he did not play a conspicuous part. If in the exercise of his humane duties, he does not always receive the plaudits of his countrymen, he never fails to obtain the approval of his own conscience, often the only reward coveted by true merit and unaffected modesty. The examples of Ambrose Paré and of Baron Larrey afford striking illustrations of the happy influence which the military surgeon is capable of exerting over the minds of soldiers in times of war, in inspiring confidence in their leaders and in their own personal safety when struck down by accident or disease. When the father of French surgery appeared at Metz, invested by the army of Charles V., the soldiers, exhausted by hunger and fatigue, crowded around the great surgeon the moment they saw him approach, exclaiming, "we have no longer any fear of dying even if we should be wounded; Paré, our friend, is among us;" and the great Napoleon declared that Larrey, who followed him through all his campaigns, was the most honest and upright man he had ever known.

In considering the contributions which have been made

by America to surgical art and science during the last hundred years, the object of this paper, it will be necessary to arrange them under different heads instead of presenting them in chronological order or historical sequence. In speaking of the honoured dead, I shall not confine myself to a mere enumeration of their labours, but append, whenever it may be convenient, brief biographical sketches, so as to place their true character more fully before the reader. Of the living I shall, for obvious reasons, say little, if anything, beyond what more immediately concerns their contributions to the general stock of American surgery. In carrying out this design it will be my earnest endeavour to do full justice to all my surgical brethren, in all sections of the country, engaged in the legitimate exercise of their profession; at the same time, however, it must be understood that the limits within which I am restricted will prevent me from entering into any minute details. It will be perceived that most of the prominent surgeons mentioned in this paper have been teachers in medical schools, and it is hardly necessary to add that they must have exercised more or less influence in moulding the surgical mind of the country. Not a few of them were the worthy peers of Roux, Lisfranc, and Dupuytren, of France; of Abernethy, Cooper, Brodie, and Lawrence, of England; of Cusack, Crampton, and Colles, of Ireland; of Bell and Syme of Scotland; of Graefe and Rust, of Germany, and of Scarpa and Porta, of Italy; men who in their respective countries stood head and shoulder in talent, influence, skill, and attainment, above most of their contemporaries.

Although this paper is designed to record the achievements of American surgeons, there are, strange to say, as a separate and distinct class, no such persons among us. It is safe to affirm that there is not a medical man on this continent who devotes himself exclusively to the practice of surgery. On the other hand, there are few physicians, even in our larger cities, who do not treat the more common surgical diseases and injuries, such as fractures, dislocations, and wounds, or who do not even occasionally perform the more common surgical operations. In short, American medical men are general practitioners, ready, for the most part, if well educated, to meet any and every emergency, whether in medicine, surgery, or midwifery. Of late, the specialists have seriously encroached upon the province of the general practitioner, and, while they are undoubtedly doing much good, it is questionable whether the arrangement is not also productive of much harm. The soundest, and, therefore, the safest, practitioner is, by all odds, the general practitioner, provided he is thoroughly educated, and fully up to his work.

The century under review opened with no great lights in practical surgery. Although the Revolutionary War had furnished a large number of army surgeons, who rendered important service both on the field and in the hospital, there was not one among them who was entitled to the term great, in the sense employed at the present day. The nearest approaches to such a distinction were Dr. William Shippen and Dr. John Warren, the first professors, respectively, of anatomy and surgery in the University of Pennsylvania and Harvard College. Dr. Benjamin Rush acted at first as Physician, and afterwards as Surgeon-General to the Middle Division of the army, and doubtless discharged well the duties of his offices; but it must be remembered that he was educated as a medical practitioner, not as a surgeon, and there is no record which goes to show that he ever performed any of the great operations of surgery. Dr. James Tilton also rendered good service; but his special province seems to have been the supervision of hospitals, and, in a general way, the care of the sick and wounded. How active and energetic he was in the performance of his varied and responsible duties, and how horoughly he had studied military hygiene, his little tract, entitled *"Economical Observations on Military Hospitals and the Prevention and Cure of Diseases incident to an Army,"* published in 1813, at Wilmington, Delaware, sufficiently attests. It is proper to state that a portion of this tract was submitted, at the request of the Secretary of War, as a report to a committee of Congress, of which Robert Morris, the eminent financier of the Revolution, was chairman, and that it met with the warm approval of Government. There is reason to believe that Dr. Tilton's long, earnest and thoughtful services were instrumental in saving many lives. He was fully aware of the pernicious influences of hospitals, and he, therefore, availed himself, whenever practicable, of open tents for the accommodation of the sick and wounded. The motto of his tract was borrowed from Homer's Iliad:—

> On mules and dogs the infection first began,
> And last the baneful arrow fixed in man.

Tilton served during our late war with England, and died at an advanced age in 1822. Delaware may justly claim him as one of her most distinguished sons.

Rush, like Tilton, distinguished himself as a hospital physician, and during the war published a pamphlet embodying the results of his observations on the diseases of soldiers; a brochure widely disseminated and of great benefit to army surgeons. This great and good man died, in 1813, at the age of sixty-eight years. Dr. John Warren, whose active life began with the Revolutionary War, in which he was a Surgeon-General, died in 1815. His career was crowded with political events, and his name is indissolubly associated with the rise and progress of medical institutions in Massachusetts.

Two surgeons, of great name and renown, flourished contemporaneously with John Warren, although they were both by many years his juniors; while two others, destined to become equally illustrious, had just entered upon their brilliant career a short time before Warren died. These men, whose names are as familiar with the profession as household words, were Philip Syng Physick, born at Philadelphia in 1768; John Collins Warren, of Boston, born in 1778; Valentine Mott, of New York, born at Glen Cove, Long Island, in 1785; and Benjamin Winslow Dudley, of Lexington, Kentucky, born at Spottsylvania, Virginia, in 1785. Out of the loins of these men have issued, either directly or indirectly, many of the great surgical practitioners of the past and present day in this country. Wright Post, a native of Long Island, a surgeon of great note in his day, was born two years before Physick, and died in 1822. He was appointed Professor of Surgery in the Medical College of New York in 1792, and performed many highly creditable operations.

John Syng Dorsey, a nephew of Physick, and a native of Philadelphia, was born in 1783. He was the author of the first treatise on surgery ever published in this country; a work in two volumes, extensively used as a text-book in our schools, and also for a considerable period in the University of Edinburgh. After having completed his foreign studies he settled in Philadelphia, where he soon acquired a large practice, and became one of the most popular men in the community. He contributed a number of valuable papers to the medical press, was the first in the United States to ligate the external iliac artery, and, at the time of his death, which occurred in 1818, before he had attained his thirty-fifth year, he was Professor of Anatomy in the University of Pennsylvania.

The name of Ephraim McDowell will be forever famous in the history of surgery as the originator of ovariotomy. Although a native of Virginia, he earned his reputation at the town of Danville, Kentucky, where he practised his profession from an early period of his life until the time of his death, in 1830, in the fifty-ninth year of his age. His medical education was acquired mainly in the University of Edinburgh, and in the lecture room of the celebrated John Bell, of whom he was a great admirer. His first case of ovariotomy occurred in 1809. He was a successful lithotomist, and commanded a large field of practice. The chaplet that should have been worn on his brow has been placed by a grateful profession upon his tomb.

One of the most extraordinary medical men whom this country has ever produced, whether we regard his great ability as a general practitioner, his skill and daring as a surgeon, or his versatility as a teacher of the different branches of medicine, was Nathan Smith, of New Haven, who, after a brilliant career, died in 1829, at the age of sixty-seven years. He was a native of Rehoboth, Massachusetts, a professor in Dartmouth, Yale, and Bowdoin colleges, and the author of a monograph entitled a *Practical Essay on Typhous Fever*. This wonderful man, a true pioneer in the cause of medical education, lectured early in life upon all the branches of medicine then taught at Dartmouth College, and enjoyed for a long time almost an unrivalled reputation as a surgeon, teacher, and general practitioner in the New England States. In his tract on fever he shows himself to have been a sagacious observer, thoroughly acquainted with the nature and treatment of the disease which he has so well described.

Dr. John Beale Davidge, a native of Annapolis, was the founder of the University of Maryland, in which he occupied for many years the chairs of anatomy and surgery. He enjoyed a high reputation as a teacher, and as a practitioner of medicine, surgery, and of midwifery. As a writer, he occasionally contributed to the periodical press; and he is said to have possessed a considerable amount of literary attainment. He died in 1829, at the age of sixty years, leaving behind him an enviable name.

George McClellan, a contemporary of Physick, Mott, Warren, and Dudley, was born at Woodstock, Connecticut, in 1796, and became distinguished early in life as a bold and dashing operator, and as a fascinating, enthusiastic, and instructive teacher. He was the founder of the Jefferson Medical College of Philadelphia, in which he was the professor of surgery from 1825 until 1838, and the author of a work entitled *The Principles and Practice of Surgery*, edited as a posthumous production by his son, the late Dr. John H. B. McClellan. McClellan died in 1847, in the fifty-first year of his age.

Jacob Randolph, a man of prominence in his day, will be remembered chiefly in connection with his efforts to popularize lithotrity in this country, an operation in which he obtained considerable reputation as a skilful manipulator. Born in Philadelphia in 1796, he was for a number of years a hospital surgeon, and for a short time Professor of Clinical Surgery in the University of Pennsylvania. He contributed some valuable papers to the periodical press, and wrote an able and graphic memoir of his father-in-law, Dr. Physick. "Dr. Randolph," writes one of his biographers, "was endowed in a high degree with all the attributes of the great surgeon." His death occurred in 1848, in the fifty-second year of his age.

William Gibson, a native of Maryland, occupied the chair of surgery in the University of Pennsylvania from 1818 until 1854, when advancing age and bodily infirmity compelled him to resign. He was an accomplished lecturer, a lucid writer, an able operator, and the author of a work entitled *The Institutes and Practice of Surgery*, extensively used as a text-book in its day. Dr. Gibson had the good fortune to be the first surgeon that ever tied the common iliac artery, an operation which contributed greatly to the extension of his reputation.

Among the surgeons who occupied a conspicuous place in this country, as teachers and practitioners, during the middle of the present century, may be mentioned the names of Thomas Dent Mütter, Daniel Brainard, David Gilbert, George Hayward, and George W. Norris. Mütter, a native of Virginia, and for fifteen years Professor of Surgery in the Jefferson Medical College of Philadelphia, distinguished himself as a polished, forcible, and popular teacher, and as an able plastic surgeon. He died in 1859, leaving his surgical museum to the College of Physicians of Philadelphia, together with $30,000 for its increase, and for the endowment of what is known as the Mütter lectureship in that institution. Brainard, born in 1812, in the State of New York, was the founder of Rush Medical College of Chicago, and for many years its professor of surgery. His essay on ununited fractures, in which he advanced some novel methods of treatment, received the prize of the American Medical Association, at its meeting at St. Louis, in 1854. A bold operator and a successful practitioner, he was an excellent teacher, an original thinker, and a good writer. He died in 1866, after having held for many years the leadership of surgery in the northwestern States of the Union. Gilbert, whose death occurred in 1868, at the age of sixty-five, was a native of Pennsylvania, and for a number of years Professor of Surgery in the Pennsylvania Medical College of Philadelphia. He was distinguished as a skilful operator, as well as an excellent general practitioner; and, although his surgical exploits were not numerous, they were of such a nature as to make his name widely known. Hayward, a contemporary and for many years a colleague of John C. Warren as Professor of Clinical Surgery at Harvard University, was one of the leaders of his profession in New England, enjoying a large reputation as an operator, and as a general practitioner. He was the first to make known among us the writings of Bichat, having translated his immortal work on general anatomy, a task which of itself entitles him to no ordinary praise. Dr. Hayward was born in 1791, and died in 1863, at the age of seventy-three years. Dr. George W. Norris, a Philadelphian by birth and education, for many years surgeon to the Pennsylvania Hospital, will be remembered chiefly by his essay on ununited fractures, and by his statistics of the results of operations upon the larger arteries, of fractures, and of amputations, published originally in the *American Journal of the Medical Sciences*, and collected a short time before his death, in 1875, in an octavo volume entitled *Contributions to Practical Surgery*. Dr. Norris was for a short time Professor of Clinical Surgery in the University of Pennsylvania. His great forte was conservative surgery, in which he achieved some of his proudest triumphs.

To the above list must be added the name of John Rhea Barton, whose operative skill placed him, a third of a century ago, in the foremost rank of his contemporaries. Although he was essentially a mechanical surgeon, he was not the less an excellent diagnostician, and an able general practitioner. To him the profession is indebted for the invention of a valuable operation for the relief of anchylosis, mentioned in another part of this paper, and also for an admirable account of fracture of the inferior extremity of the radius, known as Barton's fracture. He was also the first to suggest the use of the bran-dressing, so useful in the treatment of compound-fractures and dislocations of the leg, and he devised an excellent bandage for fractures of the lower jaw. As an expert lithotomist, he had few rivals in his day. After having been engaged for twenty years in active practice, Dr. Barton retired to private life, wealth, it is said, having allured him from his profession. He died in 1871.

Four New York surgeons deserve honourable mention in

connection with the brilliant career of Valentine Mott, of whom all were contemporaries, if not at some time or another colleagues—Richard S. Kissam, J. Kearney Rodgers, Alexander H. Stevens, and John Watson, all hospital surgeons, and two of them excellent writers, one a distinguished teacher, and all able operators and skilful practitioners. It is said of Kissam, who died in 1822, that of sixty-five cases of lithotomy which passed through his hands only three proved fatal; a degree of success eminently creditable to his skill and judgment. J. Kearney Rodgers, born at New York in 1793, a pupil of Wright Post, and afterwards of Sir Astley Cooper, Travers, Abernethy, and Sir B. C. Brodie, founded, along with Dr. Edward Delafield, the New York Eye and Ear Infirmary, and was, for many years, one of the surgeons of the New York Hospital. He was distinguished for his great practical tact, his ability as a diagnostician, and his great adroitness and elegance in the use of the knife; but, above all, for his many manly and noble traits of character. His great operation, one which made his name known throughout the world, was the ligation, in 1846, of the left subclavian artery between the scalene muscles. His death occurred in November, 1857. Alexander H. Stevens, who died only a few years ago at a ripe old age, spent in retirement at his residence at Astoria, Long Island, distinguished himself early in life by his translation of a portion of Boyer's treatise on *Surgery*, and, subsequently, by the publication of several valuable surgical memoirs; he was one of the surgeons of the New York Hospital, and for a number of years professor of surgery in one of the New York medical colleges. As a clinical teacher, he enjoyed a high reputation, and was greatly beloved for his amiable and excellent qualities. John Watson, a native of Londonderry, Ireland, where he was born in 1807, immigrated to New York at an early age, and soon became distinguished as an able practitioner, a classical scholar, an admirable writer, and an acute critic. He was for a long time surgeon to the New York Hospital, was a skilful operator, and was a copious contributor to the medical press, especially to the pages of the *American Journal of the Medical Sciences*. His *Medical Profession in Ancient Times*, published in 1856, is a highly creditable production. He had one of the largest and choicest collections of the works of the fathers of the profession in this country. Dr. Watson died at a comparatively early age.

To the above catalogue must be added the name of a distinguished surgeon, Dr. George Bushe, an Irishman, brought over to this country by the faculty of the Rutgers Medical College of New York, in 1828, as professor of anatomy in that school on the recommendation of Mr., afterwards Sir William, Lawrence. Bushe died young, leaving behind him a brilliant reputation as a bold, dashing operator, and as the author of the well-known standard monograph on the *Diseases of the Anus and Rectum*, long, if indeed not still, the ablest work on the subject in any language. He also published a memoir on staphylorraphy, and was the founder and editor of the *New York Medico-Chirurgical Bulletin*, an able journal of brief duration.

Granville Sharp Pattison, a name well known in America and Europe, was a Scotchman, and a pupil of Allan Burns, of Glasgow, the author of the great work entitled *Observations on the Surgical Anatomy of the Head and Neck*, of which he brought out an edition in this country. In 1820 he was appointed Professor of Surgery in the University of Maryland, and on retiring from that institution he occupied successively the chair of anatomy in the London University, in the Jefferson Medical College, and in the University of the City of New York. Although he was not a great surgeon, he was one of the ablest teachers of surgical anatomy of the age, and by his enthusiasm as a lecturer he had the happy faculty of inspiring his pupils with a love for surgery such as few men ever possessed. He contributed a number of valuable papers to our periodical literature, and was for some time one of the editors of the Philadelphia *Medical Recorder.* Dr. Pattison died at New York in November, 1851, aged sixty years.

Northern New England can boast of two representative surgeons, of great, if not unsurpassed, ability as practitioners and operators, both natives of New Hampshire, men of great renown, and of unsullied character. I allude to Amos Twitchell, of Keene, born in 1781, and Dixi Crosby, for many years the distinguished Professor of Surgery at Dartmouth College. Both of these men performed a vast amount of work in their day; they both possessed uncommon skill in the use of the knife; and such was the confidence reposed in their judgment, ability, and integrity, that patients flocked to them from all sections of the New England States for aid and advice, often in cases of great emergency and suffering, in which relief had been sought in vain in other quarters. Twitchell, whose life was closed in May, 1850, performed many bold and difficult operations; but the crowning glory of his life, as remarked by his biographer, Dr. Albert Smith, of Dartmouth, was the ligation of the primitive carotid artery in a case of secondary hemorrhage, a feat which he executed successfully in 1807, eight months prior to Sir Astley Cooper's famous case, supposed, until recently, to have been the first of its kind upon record. Of Dr. Crosby, whose recent death was so widely deplored as a great loss to his country and his profession, nothing further need be said here, as the following pages bear ample testimony to his skill and judgment as a great surgeon.

Thomas Hubbard and Jonathan Knight, of New Haven; Alban Goldsmith, of New York; Horatio G. Jameson, of Baltimore; John Wagner, James Ramsay, and John Bellinger, of Charleston; Joshua B. Flint, of Louisville; P. C. Spencer, of Petersburg; Joseph Parrish, Joseph Hartshorne, Thomas T. Hewson, William E. Horner, and Thomas Harris, of Philadelphia, were well-known surgeons in their day, distinguished either as teachers, as operators and practitioners, or as contributors to our periodical literature. The popularity of Professor Knight, as a gentleman and an honoured member of the profession, was shown by his having been twice elected President to the American Medical Association, a compliment never before nor since accorded to any of its members. Dr. Hewson, a son of the celebrated London anatomist, was for nearly a third of a century a hospital surgeon. Dr. Harris was a naval surgeon, who late in life held the office of Chief of Bureau at Washington City. Among those who closed their earthly career, either quite recently or at a comparatively recent period, stand conspicuously the names of Horace A. Ackley, Reuben D. Mussey, and George C. Blackman, of Ohio; of Charles A. Pope, of Missouri; of J. Tyler Bradford, and James M. Bush, of Kentucky; of Thomas B. Buchanan, of Tennessee; of Warren Stone, of Louisiana; of Josiah C. Nott, of Alabama; of Charles Bell Gibson, Hugh Holmes McGuire, and John P. Mettauer, of Virginia; of Paul B. Goddard, John H. B. McClellan, and George W. Norris, of Pennsylvania; of Alden March, Ernst Krackowizer, and James A. Armsby, of New York; and J. Mason Warren, and Winslow Lewis, of Massachusetts.

It is not probable that America will ever again produce four surgeons of equal renown with Philip Syng Physick, John C. Warren, Valentine Mott, and Benjamin W. Dudley, for the reason that it is not at all likely that an equal number of young practitioners will ever again be placed under equally advantageous circumstances for their development. When Physick commenced the practice of medicine in Philadelphia there was no surgeon of any prominence north of New York, and very few of any distinction even on that side of the line. He returned after an absence of four years with the prestige of foreign study, as a favourite pupil of the celebrated John Hunter, and he had hardly touched his native shore before there was a frightful outbreak of yellow fever in his own city, carrying off not less than 4000 of its inhabitants in 1793. Physick was at once appointed physician to the Bush Hill Fever Hospital, where, as well as in the city, then comparatively small, he was brought into contact with many prominent citizens, a circumstance highly favourable, one would suppose, to his speedy introduction into practice; and yet practice came so slowly that he for some time seriously contemplated abandoning the profession and settling upon a farm. Gradually, however, this chilling and discouraging feeling, so often experienced by the young aspirant after fame, wore off, and long before he had reached the meridian of life, he stood at the head of his profession as the first surgeon on the American continent. His appointment to the chair of Surgery in the University of Pennsylvania, in 1805, greatly promoted his interests, and was a means of attracting patients to him from all parts of the country. Upon the death of John Syng Dorsey in 1818, Physick was transferred from the surgical to the anatomical chair, which he occupied until 1831, when advancing age and increasing physical infirmities compelled him to retire.

The Father of American surgery, a title well deserved because well earned, died in 1837, in the sixty-ninth year of his age, having only a short time before that event performed several important operations, among others that of lithotomy upon the venerable Chief Justice Marshall, from whose bladder he removed upwards of one thousand calculi. Physick has left no works to commemorate his fame or to record his vast experience, a few short papers in the medical press of the day comprising the whole of his contributions to the surgical literature of the country. In Dorsey's *Elements of Surgery* may be found an abstract of the experience of the first twenty-five years of his surgical practice. Physick was deeply imbued with the doctrines of his illustrious preceptor, John Hunter, and constantly advocated their importance in his lectures, thus contributing, in no small degree, to their dissemination and appreciation in this country, and, more or less directly, to the advancement of American surgery. As a surgeon, it may be truly said of Physick, what Dr. John Brown has said of Mr. Syme, that he never spilt a drop of blood uselessly, or, as a teacher, ever wasted a word.

John C. Warren was a son of John Warren, the first professor of anatomy and surgery in Harvard College, and a nephew of General Joseph Warren, who lost his life at the battle of Bunker Hill. On the death of his father, in 1815, he succeeded to his two chairs, which he occupied for upwards of a quarter of a century; he was for a long time surgeon to the Massachusetts General Hospital, and there was no medical man in the New England States who was so favourably known, or who commanded so wide an influence as an operator and general practitioner. From his aristocratic descent and his official relations, it is evident that John C. Warren had a sort of pre-emption right to the surgery, not only of Massachusetts, but of all the surrounding States. He was the author of a work on Tumours, was a large contributor to the periodical press, and was the first who ever administered ether as a preventative of pain in a surgical operation.

Valentine Mott, in the thirty-fourth year of his age, tied the innominate artery, a feat never accomplished before, and on waking up the next morning found himself "the observed of all observers." Other great operations followed in more or less rapid succession, and fame soon perched upon his brow, carrying his reputation into all parts of the civilized world. When Mott, after his return from Europe, settled in New York, his only surgical competitors, of any note, were Wright Post and Richard S. Kissam, good but not great surgeons. He had thus an open field, which he long successfully occupied, although latterly not without many able competitors, and even rivals, up to the time of his death in 1865.

B. W. Dudley, in 1810 went to Europe, where he availed himself of the instructions of Cline, Abernethy, and Cooper, in London, and of Boyer, Dubois, Larrey, and others in Paris. Bringing back with him French manners, which he affected during the remainder of his life, he settled at Lexington, Kentucky, then a small village, in 1814. Upon the organization of Transylvania University in 1819, he was appointed professor of anatomy and surgery, the latter position of which he held until the school was finally closed in February, 1850. Dudley early in his career had no competitor. Ephraim McDowell, a resident of Danville, Kentucky, only 36 miles from Lexington, had, it is true, already performed ovariotomy several times as well as many other important operations, but the former of these feats, instead of enhancing his reputation, only served to bring him into ridicule, if not positive contempt, both at home and abroad. In all the great West and Southwest there was not one surgeon of commanding skill, talent or reputation. A great field here lay fallow, and Dudley soon, of necessity, became its occupant and its successful cultivator; so true is it that circumstances more frequently make men than talent and genius, or great and intrinsic merit. Dudley was a great advocate of protracted rest and low diet in the treatment of chronic inflammation, and of the bandage as a means of controlling swelling and muscular action in the treatment of wounds, fractures, and dislocations. Indeed, he might be said to have been the knight of the roller, so generally did he employ it, and so strongly advocate its utility. His disciples, less skilled in its application, of course committed many egregious blunders with it, causing much suffering with the occasional loss of a limb, and a suit for malpractice.

Dudley expired in January, 1870, in the eighty-fifth year of his age, most of his time, since he delivered his last lecture, in 1850, having been spent in retirement, in a species of gradually increasing imbecility.

In connection with these men, who were the surgical autocrats of their day in this country, must be mentioned the name of Dr. Warren Stone, who, in point of reputation and professional pre-eminence, occupied the same position in the Southern States that they, respectively, did in the Eastern, Northern, and Western. Born at St. Albans, Vermont, February 8th, 1808, Stone settled early in life at New Orleans, where he soon acquired a degree of popularity seldom equalled in any walk of life. A man of talents and of wonderful kindness and benevolence, he was an attractive talker, and a boon companion, with a smile and a cordial shake of the hand for every one who approached him. The boatmen of the Mississippi and Ohio Rivers literally worshipped him. As he grew in reputation, nobody could be sick without having Stone, either as attending or consulting surgeon or physician. He was the great commoner in his day in the South; tall of stature, not particularly refined or elegant in his address, but so kind and winning in his manners as to inspire his patients with unbounded confidence in his ability and skill. The great secret of his success lay in his large heart and in the native powers of his mind, strong and well poised, but not at all cultured. When he passed away, in December, 1872, the Southern people mourned his loss as the loss of a household god. Stone has left no substantial memorial of his labours. His vast experience, as a surgeon and physician, is buried with his ashes. Authorship had no charms for him. He was not a great, much less a brilliant, operator; and, as a teacher of surgery, he was too erratic and too unsystematic to do justice to the chair which he held for a third of a century in the University of Louisiana. When he settled at New Orleans the only surgeon of any note was Dr. Charles A. Luzenberg, a man of elegant manners, an excellent scholar, and a brilliant practitioner, as well a dexterous operator, who died in 1847, thus leaving the field to his young rival, who had already for several years past been treading closely upon his heels.

All these men, and many others equally good, although not equally distinguished, have passed away. Had the lot of the very foremost of them been cast in our day, they would have had many competitors, and not a few successful rivals in the race of fame. In short, they never would have attained such wondrous pre-eminence. If circumstances did not make them what they were, circumstances powerfully contributed to their development, and in giving saliency to their character. None of them were brilliant; none even uncommonly talented. The men now upon the stage have nothing to be ashamed of; educated more or less thoroughly, they are fully equal to their work, and are, in every respect, worthy successors in an age when science and skill occupy a much more exalted position than they did in the days of the Father of American Surgery.

In the *Surgery of the Bloodvessels*, America need not be ashamed of her achievements, of which some have certainly been eminently daring and brilliant. Commencing with that on the vessels of the neck, it may be stated that the common carotid arteries have been ligated in innumerable instances, both on account of hemorrhage in wounds of the cervical region and in their continuity for the cure of aneurism and of morbid growths. The first operation in which, on this continent, the primitive carotid artery was secured with a double ligature was performed in 1803, by Dr. Mason Fitch Cogswell,[1] of Hartford, Connecticut, the procedure having been rendered necessary during the extirpation of a "scirrhous tumour" of the neck, in which that vessel was

---

[1] N. E. Journ. of Med. and Surg., vol. xiii. p. 357.

deeply embedded. The ligature came away at the end of two weeks, and the man lived until the twentieth day, when he died exhausted from general debility, hastened by slight bleeding from a small vessel near the angle of the jaw. Dr. Cogswell, who was an army surgeon in the Revolutionary War, died in 1830. In its continuity, the artery, if I am not in error, was first tied in this country in 1813, by Wright Post,[1] of New York, in a case of aneurism, followed by the recovery of the patient. To Dr. Macgill,[2] of Maryland, belongs the credit of having been the first to secure successfully both carotids, after an interval of one month, on account of a fungous growth in the orbit of each eye, the operations having been performed in 1823, four years after that of Dupuytren and Robert. Altogether, this procedure has been executed fifteen times on this side of the Atlantic, with the result of 11 recoveries, 2 deaths, and 2 failures. The operators were Macgill, Mussey, Mott, F. H. Hamilton, John Ellis, J. M. Warren, George C. Blackman, Reynolds and Van Buren, Willard Parker, J. R. Wood, J. C. E. Weber, J. M. Carnochan, and H. E. Foote. The caused necessitating the operations were, for the most part, epilepsy, erectile tumours, or malignant growths of the orbit of the eye. In Carnochan's[3] case it was performed for the relief of elephantiasis of the neck, face, and ear. In one of Mott's cases,[4] where the interval of the application of the two ligatures was only fifteen minutes, death ensued within twenty-four hours. In Dupuytren and Robert's case the interval was thirty-six years, a sufficient reason for excluding it from the list, as the parts had long ago accommodated themselves to the changes induced in the cerebral

---

[1] Am. Med. and Phil. Register, vol. iv. p. 366.
[2] N. Y. Med. and Phys. Journ., vol. iv. p. 576.
[3] Am. Journ. Med. Sci., July, 1867, p. 109.
[4] Mott's Velpeau, edited by Blackman, vol. i. p. 867.

circulation by the first operation. Mott tied the common carotid artery altogether fifty-one times, in most of the cases successfully. During my pupilage in this city, in 1827, I assisted the late Dr. George McClellan in ligating this vessel in a child only five months old, on account of an immense nævus on the upper part of the face. The descending branch of the ninth pair of nerves was divided in the operation, as it interfered with the passage of the ligature. The infant speedily recovered, without, however, any material benefit as it respected the morbid growth. Mott,[1] in a similar case, tied the artery successfully, followed, it is said, by a cure of the aneurism, in a child only three months of age. A case was reported in 1857 by Dr. Gurdon Buck,[2] of New York, in which, on account of a deep wound of the parotid region, a ligature was successfully applied by that gentleman simultaneously to the common and internal carotid arteries. Two examples of a similar kind have occurred since that period: one, in 1871, in which Professor W. T. Briggs,[3] of Nashville, tied both these vessels, on account of secondary hemorrhage after an operation for the cure of a traumatic aneurism of the common carotid artery; and the other, in 1872, in which they were secured by Professor A. B. Sands,[4] of New York, on the tenth day after excision of the left half of the lower jaw. Both patients recovered. In September, 1875, Dr. Donald Maclean, Professor of Surgery in Michigan University, Ann Arbor, in a case of traumatic aneurism, cut down upon the tumour, and, turning out the clots, tied the common carotid at both ends.

The innominate artery was approached for the first time with a ligature on the 11th of May, 1818, the operator being

[1] Am. Journ. Med. Sci., Nov. 1830, p. 271.
[2] Ibid., Jan. 1856, p. 267.
[3] Nashville Journ. Med. and Surg., March, 1871, p. 103.
[4] New York Med. Journ., Jan. 1874, p. 34.

Dr. Valentine Mott. The patient was Michael Bateman, fifty-seven years of age, the subject of an aneurism of the right subclavian artery, the tumour being of large size, well marked, and the seat of much suffering. The artery was tied about half an inch below its bifurcation; the ligature was detached on the fourteenth day; on the ninth day there was some hemorrhage, and again on the twenty-third, but in larger quantity; and death occurred on the twenty-sixth day from exhaustion. The dissection revealed absence of occlusion, and extensive ulceration of the structures at the lower part of the neck, involving the innominate artery. Dr. Richard Wilmot Hall,[1] of Baltimore, repeated the operation in 1830; in 1859 it was performed by Dr. E. S. Cooper, of San Francisco;[2] and in 1864 by Dr. A. W. Smyth,[3] of New Orleans, who at the same time ligated the common carotid artery. In this case, which finally, after repeated attacks of hemorrhage, terminated successfully, the fortunate result was manifestly due to the ligation of the vertebral artery on the fifty-fourth day after the primary operation. The particulars of Dr. Mott's case will be found in the *New York Medical and Surgical Register* for 1818, and also in Townsend's translation of Velpeau's Operative Surgery. This operation, which reflects imperishable credit upon Dr. Mott, as a skilful and daring surgeon, was performed before he had been thirteen years in active practice. Having made himself thoroughly familiar with the surgical anatomy of the neck, he had no hesitation in attempting it, satisfied that he possessed the requisite courage, judgment, and dexterity to complete it. The case of Dr. Smyth is replete in interest, not only as illustrative of extraordinary ability of the operator, but as showing how recovery may occasionally occur

---

[1] Baltimore Med. and Surg. Journ., vol. i. p. 125.
[2] Am. Journ. Med. Sci., Oct. 1859, p. 395.
[3] Ibid., July, 1866, p. 281.

under, apparently, the most desperate circumstances. It is proper to add, that, in all the other cases, amounting to upwards of a dozen, in which the innominate artery was tied, the result was unfavourable, the immediate cause of death being secondary hemorrhage.

The subclavian artery has been repeatedly secured on the capular aspect of the scalene muscles, both for the arrest of hemorrhage and the cure of aneurism. The first successful operation for the cure of aneurism, in this country, if not in the world, was performed by Dr. Wright Post[1] in 1817. The credit arising from the case is greatly enhanced by the fact that the operation, a very delicate one, had previously failed in the hands of such " master spirits in surgery," as Ramsden, Abernethy, and Cooper. Ligation of the subclavian artery on its tracheal aspect, originally executed by Mr. Colles, of Dublin, has, I believe, been performed only three times in this country, the surgeons being Valentine Mott,[2] J. Kearney Rodgers,[3] and Willard Parker[4] Until the operation was done by Rodgers, such an attempt was universally regarded as impracticable on the right side for the relief of aneurism, from the close proximity of the vessel to the sac of the pleura and the intimate relations of the tumour with the thoracic duct and the great vessels and nerves of the neck. The patient succumbed under the effects of secondary hemorrhage on the fifteenth day. In 1863, Professor Parker,[5] also in a case of aneurism, performed a similar operation, at the same time securing the common carotid and vertebral arteries, in the hope of thus effectually preventing the occurrence of secondary hemorrhage. Despite

---

[1] Trans. N. Y. Phys.-Med. Soc., vol. i. p. 387.
[2] Am. Journ. Med. Sci., Aug. 1833, p. 354.
[3] Ibid., April, 1846, p. 541.
[4] Ibid., April, 1864, p. 562.
[5] Am. Med. Times, March 5, 1864, p. 114.

of this precaution, however, the patient died from this cause at the end of the sixth week. There are two cases, one in 1867 by J. C. Hutchison,[1] of Brooklyn, and the other in 1868 by A. B. Sands,[2] of New York, in which a ligature was placed in immediate succession upon the common carotid and subclavian arteries for aneurism of the innominate. An instance in which the carotid alone was secured for a similar disease occurred in 1867, in the hands of Dr. Addinell Hewson,[3] of Philadelphia. Statistics go to show that ligation of the carotid alone is generally a more rapidly fatal operation than the simultaneous ligation of this artery and of the subclavian. Dr. Thomas G. Morton[4] in 1866, in a case of spontaneous axillary aneurism at the Pennsylvania Hospital in a man fifty-one years of age, tied the left subclavian artery between the scalene muscles, the patient finally recovering, after amputation of the limb at the shoulder-joint, rendered necessary by sloughing of the tumour, followed on the forty-third day by secondary and frequently recurring hemorrhage, and eventually by mortification of the extremity.

Professor Pancoast, many years ago, suggested a more easy method than the one usually adopted for tying the subclavian artery below the clavicle, particularly applicable to cases of aneurism of the neck reaching so far down as to allow little space for exposing the vessel. The operation consists in opening the fissure between the sternal and clavicular attachments of the great pectoral muscle, when the former is cut across immediately below the collar-bone, and the artery is sought for and ligated.

Sir Astley Cooper, in 1817, executed the daring and brilliant feat of ligating the abdominal aorta in a man thirty-

---

[1] Med. Record, vol. 2, p. 265.
[2] Ibid., vol. 3, p. 531.
[3] Penn. Hosp. Rep., vol. i. p. 219.
[4] Am. Journ. Med. Sci., July, 1867, p. 70.

eight years old, on account of an aneurism of the left iliac artery; and, although the case terminated fatally, it has induced a number of other surgeons to follow his example. In this country the operation was performed for the first and only time, in 1868, by Professor McGuire,[1] of Richmond, Virginia, the patient being a man thirty-six years of age, the subject of aneurism of the external and common iliac arteries, involving the lower portion of the aorta. Death occurred within eleven hours after the operation.

Ligation of the common iliac artery was first practised in 1812, by Dr. William Gibson,[2] of Baltimore, afterwards Professor of Surgery in the University of Pennsylvania, for the arrest of hemorrhage caused by a gunshot wound of the abdomen. Death occurred on the fifteenth day, from gradual loss of blood. The first case in which the operation was performed successfully for the cure of aneurism was that of Dr. Mott[3] in 1827. Among American surgeons who have tied this vessel for aneurism may be mentioned the names of Charles A. Luzenberg, Edward Peace, Warren Stone, A. J. Wedderburn, W. A. Van Buren, Stephen Smith, L. A. Dugas, Alban Goldsmith, William Hammond; and for the arrest of hemorrhage, those of Alfred Post, Willard Parker, and Gurdon Buck.[4] In the latter case, one of aneurism of the femoral artery, ligatures were successively applied to the femoral, profunda, external iliac, and common iliac. In Dr. Stone's[5] case, fatal on the twenty-sixth day, the artery was included in a silver wire ligature, probably the first instance of the kind on record. In a case in the practice of the late Dr. George Bushe,[6]

---

[1] Am. Journ. Med. Sci., Oct. 1868, p. 415.
[2] Am. Med. Recorder, vol. 3, p. 185.
[3] Phila. Journ. of Med. and Phys. Sci., vol. xiv., p. 176.
[4] Am. Journ. Med. Sci., July, 1860, p. 24.
[5] Ibid., Oct. 1859, 570.
[6] N. Y. Med.-Chir. Bulletin, vol. i. p. 55.

Professor of Anatomy in Rutgers Medical College, New York, the right common iliac artery was succesfully tied in a child only six weeks old on account of extensive telangiectasis of the perineum, genital organs, anus and rectum. The difficulties of such an operation, at so tender an age, were immense, and could only have been surmounted by the most consummate skill.

The internal iliac artery was first successfully tied in 1847 by S. Pomeroy White,[1] of Hudson, afterwards of New York, for gluteal aneurism. Among other operators have been V. Mott, J. Kearney Rodgers, H. J. Bigelow, Gilman Kimball, and Thomas G. Morton.

The external iliac artery was tied for the first time in the United States in 1811 by John Syng Dorsey,[2] in a case of inguinal aneurism, the patient making a good recovery. It was the eighth operation of the kind ever performed, the first having occurred in the hands of Mr. Abernethy in 1796. Ligation of the sciatic artery for aneurism of this vessel, and subsequently, on account of secondary hemorrhage, of the common iliac, has been practised by L. A. Dugas,[3] of Georgia. Death followed on the fourth day. In a case of ligation of the external iliac, the aneurism adhered so firmly to the peritoneum that the operator, Wright Post,[4] of New York, in order to separate it, was compelled to cut through that membrane, notwithstanding which the patient made an excellent recovery.

The gluteal artery has been tied twice in this country for the cure of aneurism; first by Dr. John B. Davidge, of Baltimore, and, secondly, by Dr. George McClellan[5] of

---

[1] Am. Journ. Med. Sci., Feb. 1828, p. 304.
[2] Eclectic Repertory, vol. ii. p. 111.
[3] Southern Med. and Surg. Journ., Oct. 1859, p. 651.
[4] Am. Med. and Phil. Reg., April, 1814, p. 443.
[5] Mott's Velpeau, edited by Blackman, vol. i. p. 795.

Philadelphia. Both patients recovered, although they had lost much blood.

In my memoir of Dr. Mott, published in 1868, occurs the following paragraph in relation to the ligation of arteries by this wonderful surgeon:—

"No surgeon, living or dead, ever tied so many vessels, or so successfully, for the cure of aneurism, the relief of injury, or the arrest of morbid growths. The catalogue, inclusive of the celebrated case of the innominate artery, comprises eight examples of the subclavian artery, fifty-one of the primitive carotid, two of the external carotid, one of the common iliac, six of the external iliac, two of the internal iliac, fifty-seven of the femoral, and ten of the popliteal; in all one hundred and thirty-eight."

The first successful cure of aneurism by digital compression occurred in 1847, in the practice of the late Professor Knight,[1] of New Haven. The case was one of popliteal aneurism. In two instances of subclavian aneurism, in the hands of J. Mason Warren,[2] too far advanced for ligation, a cure was effected by direct compression of the sac, aided by the application of bags of ice.

Until a very recent period the idea was very common among surgeons, even the most enlightened and experienced, that the ligation of a vein, especially a large one, was almost uniformly productive of very grave consequences, occasionally followed by death. How utterly unfounded this opinion is, has been abundantly proved by the able and exhaustive statistical paper of Dr. S. W. Gross, of Philadelphia, published in the January and April numbers of the *American Journal of the Medical Sciences* for 1867.

The *metallic ligature* for the ligation of arteries, an American device, is much less frequently employed than, in my opinion, it deserves to be. Such a substance can, of course,

[1] Am. Journ. Med. Sci., July, 1848, p. 255.
[2] Surgical Observations with Cases, Boston, 1867.

never take the place of the ordinary ligature in the case of the smaller arteries, but for the larger trunks nothing could possibly be more eligible, especially when an operation is performed for the cure of aneurism, in which it is always very desirable to avoid suppuration, an occurrence which is almost inevitable when the common ligature is used from its tendency to act as a seton or foreign body. The metallic ligature is a non-irritant, and, if properly applied, is sure, in a sound artery, to become speedily encysted, remaining thus ever afterwards as a harmless tenant. This never happens with the ordinary ligature, which therefore never fails to keep up discharge until it is detached, whether its retention be short or long. The innocuous character of the metallic ligature was first satisfactorily demonstrated by the late Dr. Henry S. Levert, of Mobile, Alabama, while a student in Philadelphia, in a series of well-conducted experiments upon the inferior animals, performed at the suggestion of Dr. Physick, the results being embodied in his inaugural dissertation which was afterwards published, by order of the Faculty of the University of Pennsylvania, in the *American Journal of the Medical Sciences* for 1829. So far as my information extends, the late Dr. Warren Stone,[1] of New Orleans, was the first to apply a wire ligature to a human artery. The case, which occurred in 1859, was one of aneurism of the external iliac, for which he tied the common trunk of that vessel. In 1866, a similar operation was performed by Dr. C. H. Mastin,[2] of Mobile, upon the external iliac, for an inguinal aneurism; and about the same time I secured that vessel for a similar purpose. Since then I have applied the silver wire ligature to other arteries, and have had every reason to be satisfied with the results of the treatment. If surgeons only knew, or knowing, considered the advantages

---

[1] Am. Journ. Med. Sci., Oct. 1859, p. 570.
[2] Ibid., Oct. 1866, p. 580.

of the metallic ligature, I feel confident that it would be much more frequently, if not generally, employed. An account of a series of interesting experiments upon metallic ligatures, performed by Dr. Benjamin Howard, will be found in the New York *Medical Record* for 1868, and a paper illustrative of the same subject, by Dr. F. D. Lente, of Cold Spring, New York, in the *American Journal of the Medical Sciences* for April, 1869.

Dr. A. M. Pollock, of Pittsburgh, in 1859, conceived the idea of employing the wire loop as a substitute for the ligature, and he has adopted the method with great success in a number of amputations and other operations. The chief advantages which he claims for this treatment are, the more frequent union of the wound by the first intention, less danger of secondary hemorrhage, equal facility of application, and removal at the pleasure of the operator. The first case in which the wire loop was used by Dr. Pollock occurred in January, 1860. In an article in the *New York Medical Journal* for July, 1869, he has given an account of twenty-six amputations, in which this procedure was adopted, including forty-seven arteries, of which seventeen were femoral. Ingenious contrivances for suppressing hemorrhage have been devised by Professor N. R. Smith, of Baltimore, and Dr. S. F. Speir,[1] of Brooklyn. Of the intrinsic value of these different methods of treatment, it would be premature to attempt to form an estimate, as they have not been sufficiently tested, or sanctioned by the experience of the profession.

The practice of employing *animal ligatures* originated with Dr. Physick early in the present century, under the conviction that they would occasion much less irritation than ordinary ligatures, then, and still so much in use. The substance which he selected for the purpose was buckskin,

---

[1] Med. Record, April, 1871.

cut into suitable strips, which were then rolled upon a marble slab to impart to them the requisite degree of hardness, roundness, and smoothness. Dr. Dorsey, after numerous experiments with various kinds of animal substances, performed at the instance of Physick,[1] was induced to give the preference to French kid, divested of its coloured and polished surface; and such was his confidence in the safety of this material, when properly prepared, that he employed it in various amputations, and in a number of capital operations, always cutting off the ends close to the knot, and treating the wound as if no ligature had been used. He found that in the course of a few days the ligature was completely, or almost completely, dissolved in the wound without any detriment to the artery. Dr. Hartshorne, of Philadelphia, soon after, in a case of amputation of the thigh, tied up the bleeding vessels with strips of parchment. Dr. Horatio G. Jameson,[2] of Baltimore, at a later period, employed the buckskin ligature. He found, as the result of his experiments upon dogs, sheep, and other animals, that, if properly managed, it soon becomes surrounded with a cyst or capsule, which itself finally disappears through the agency of the absorbents. Dr. John Bellinger, of Charleston, and Professor Eve, of Nashville, have made more or less extensive use of the sinew of the deer. Whether the animal ligature has fallen into merited neglect, I will not stop to inquire; certain it is that it is seldom employed at the present day. I have myself always preferred the ordinary silk thread, well waxed, and firmly applied.

For taking up deep-seated arteries, when accidentally divided, Physick,[3] in 1794, suggested the use of a pair of

[1] Eclectic Rep., vol. vi. p. 389, and Dorsey's Surgery, 2d ed., vol. i. Phila. 1818.
[2] Medical Recorder, January, 1827.
[3] Dorsey's Surgery, 2d edit., vol. ii. p. 182.

curved forceps, holding in its jaws a short curved needle, armed with a silk ligature. He was led to this idea by the difficulty which he experienced in throwing a ligature around the internal pudic artery, a vessel which he had the misfortune to divide in his first case of lithotomy, performed with the gorget. A useful instrument for taking up deep-seated arteries in their continuity, as in the operations for aneurism, was devised, many years ago, by the late Drs. Parrish, Hewson, and Hartshorne of Philadelphia, and is usually known by their names.

Most surgeons of the present day are agreed that almost the only safe operation for the cure of *Varicose Veins* is subcutaneous ligation. The substance commonly preferred for this purpose is the metallic ligature, first employed, if I mistake not, nearly at the same time by Dr. Richard J. Levis, of Philadelphia, and Dr. Nathan Bozeman, of New York, the former fastening the wire with a twist or knot, and the latter with his well-known button. Professor Eve, of Nashville, prefers the animal ligature, being of opinion that it is less likely to cause irritation than any other substance. He employs the same material in the treatment of varicocele. When this affection is accompanied by extraordinary elongation of the scrotum, he first retrenches the parts and then ligates the enlarged and tortuous vessels. With this view, after having pushed the testicle up to the inguinal ring, he seizes the redundant skin with a long, narrow pair of fenestrated forceps, and cuts it away with one sweep of the bistoury, taking care not to expose the vaginal tunic. The edges of the wound being transfixed with a number of pins, placed at suitable distances, the forceps are removed, when the enlarged veins are separated from the spermatic artery and deferent tube, and included in one animal ligature, drawn sufficiently tight to arrest the circulation. The wound is now closed with twisted sutures, made by fastening the pins, of which there are generally from six to a dozen, with threads passed around each in the form of the figure 8.

In connection with the subject of varicocele, it may be mentioned that one reason, probably, of the exemption of the right spermatic vein from this affection, is the presence at the opening of this vessel into the inferior cava of a distinct valve, first described by Dr. John H. Brinton,[1] of Philadelphia, no such arrangement existing on the left side.

In the treatment of *Fractures* of the long bones, we are, it may fairly be assumed, decidedly in advance of every other nation. One of the first improvements introduced into practice early in the present century was the modification of Desault's splint by Physick for the treatment of fractures of the leg and thigh. The original splint, as is well known, reached only to the level of the crest of the innominate bone. Physick, discerning its defective construction, prolonged the outer splint to the axilla, giving the upper extremity the form of a crutch, and inserting two mortise holes for the reception of the counter-extending band. The improvement thus made was marked. Within the last sixteen years a valuable addition, suggested by Dr. H. Lenox Hodge,[2] of Philadelphia, has been made to the long splint, consisting of a bar of wrought iron, furnished with movable bolts, and bent to the right or left, in accordance with the seat of the fracture. To the bar a long broad strip of adhesive plaster stretched along the front and back of the trunk, and arranged in a loop above, is fastened, and this, in turn, is secured to the chest by three horizontal bands. Thus constructed the apparatus is, probably, as perfect as any contrivance of the kind can possibly be. However this may be, it has now given way, in great measure, if not entirely, both in private and in hospital practice, to the admirable

---

[1] Am. Journ. Med. Sci., July, 1856, p. 111.
[2] Ibid., April, 1860, p. 565.

mode of treatment introduced in 1861 by Dr. Gurdon Buck,[1] of New York, in which long splints are entirely dispensed with, and the extension made with adhesive strips, fastened to the leg, and secured below the sole of the foot to a cord, playing over a pulley, and controlled by a bag of shot or other suitable contrivance, weighing from five to fifteen pounds, according to the age of the patient. The counter-extension is made with India-rubber tubing passed round the groin and perineum, and attached to the head of the bedstead.

The anterior splint, as it is named, of Professor N. R. Smith,[2] of Baltimore, as a convenient and useful contrivance in the treatment of certain kinds of fractures of the leg and thigh, is well known, not only on this side of the Atlantic but in Europe. It is especially valuable in the management of compound fractures, and did excellent service during the late war on both sides of the line. A modification of Professor Smith's splint, much employed by our Western confreres, was devised some years ago by Dr. Hodgen,[3] of St. Louis. An admirable splint, provided with a movable foot-board, and constructed upon the principle of the double inclined plane, was invented by Professor N. R. Smith[4] early in his professional life, and is well adapted to cases of fracture of the leg and thigh, admitting of the suspension of the limb. The treatment of fractures of the lower extremity, in which the counter-extension is made by the weight of the body by raising the foot of the bedstead, originally suggested, I believe, by Dr. James L. Van Ingen, of Schenectady, is now much employed by American surgeons, and often answers where the more ordinary means fail.

[1] Bull. N. Y. Acad. of Med., vol. i. p. 181.
[2] Maryland and Va. Med. and Surg. Journ., Jan. and March, 1860.
[3] St. Louis Med. and Surg. Journ., Jan. 1864.
[4] Baltimore Med. and Surg. Journ., vol. i. p. 13.

In compound fractures of the leg, the bran-dressing, introduced by the late Dr. John Rhea Barton,[1] of Philadelphia, is an extremely valuable improvement, not only as affording a comfortable lodgment for the affected limb, but also as a means of preventing the contact of flies, and the deposit and formation of larvæ, so common in hot weather when this precaution is neglected. The bracketted splint, now so much employed in the treatment of compound fractures of the lower extremity, originated, if I mistake not, with Dr. A. Hays, of Indiana, who found it very useful in cases of gunshot wounds of the leg and thigh during our war with Great Britain in 1812.

One of the most valuable improvements, purely American in its origin, introduced into the treatment of fractures, especially of fractures of the lower extremity, is the use of adhesive plaster as a means of extension and counter-extension, as well as of adjustment of the ends of the fragments. The first public notice of this method of treatment appeared in my work on the *Diseases and Injuries of the Bones and Joints*, published in this city in 1830. I had witnessed the beneficial effects of this mode of making extension in a case of complicated fracture of both bones of the leg in the hands of my first preceptor, the late Dr. Joseph K. Swift, of Easton, Pennsylvania, and I subsequently employed it in my own practice. Since that period the application of adhesive plaster, as a means of making extension and counter-extension, has become generalized, and it is seldom that it is entirely dispensed with in any case of fracture of the long bones of the lower extremity. Upwards of twenty years ago, in a communication in the *Philadelphia Medical Examiner*,[2] I called attention to the claims of priority for Dr. Swift in this mode of treatment, and at the same time stated

---

[1] Am. Journ. Med. Sci., May, 1835, p. 31.
[2] Nov. 1852, p. 685.

that I had found in adhesive plaster an admirable dressing in fractures of the clavicle, ribs, and scapula. Notwithstanding this, Dr. Sayre, of New York, is usually credited with its origination.

The apparatus of the late Dr. Thomas Bond,[1] of this city, consisting of two splints, one of peculiar construction, stretched along the bones of the forearm, and provided with a knob for the accommodation of the hollow of the hand, is decidedly the best and most convenient contrivance that has ever been invented for the successful treatment of fractures of the inferior extremity of the radius or radius and ulna. The apparatus of Dr. George Fox, of Philadelphia, for the treatment of fractures of the clavicle did, for a long time, good service both in private and hospital practice, acting as a most valuable substitute for the complicated contrivance known as Desault's dressing. Of the many modifications of this apparatus, no particular mention need here be made, the most important, perhaps, being those of E. Bartlett, Richard J. Levis, and F. H. Hamilton.

The contrivances of John Rhea Barton and William Gibson for the treatment of fractures of the lower jaw, long maintained their place in the esteem of American surgeons, and were greatly in advance, in point of simplicity and efficiency, of those of European practitioners. Upon these contrivances, the apparatus of Dr. F. H. Hamilton is a decided improvement. The interdental splint, as it is called, devised almost simultaneously by Dr. Gunning, of New York, and Dr. Bean,[2] of Georgia, is, I believe, a purely American invention.

Of fracture apparatus, fracture boxes, fracture beds, and fracture chairs, the fertile genius of American surgeons has furnished an abundant supply, much that is worthless, and

[1] Am. Journ. Med. Sci., April, 1852, p. 566.
[2] Richmond Med. Journ., Feb. 1866.

much also that is eminently useful. The fracture beds of Jenks, Daniels, B. H. Coates, Addinell Hewson, and E. Cutter, are especially worthy of commendation, as is also the fracture chair of William H. Pancoast.

In the treatment of ununited fracture, the seton originally used in 1802, by Dr. Physick,[1] enjoyed, for a long time, a world-wide reputation; and, although it is now less extensively employed than formerly, it is still, in many cases, an admirable remedy, worthy of all the praise once bestowed upon it. The method of perforating the ends of the fragments with a peculiar instrument, introduced in 1852, by Dr. Daniel Brainard,[2] of Chicago, may occasionally be beneficially employed. Connecting the fragments together with an iron screw, as practised by Professor Joseph Pancoast in 1857, and since by him and other surgeons, is an eminently ingenious device, deserving of the highest commendation on account of its efficiency and freedom from danger, especially in ununited fractures of the femur and of the humerus. After excision of the ends of the fragments in this form of accident, it is sometimes expedient to unite the raw extremities with wire, an operation originally suggested, I believe, by Horeau in 1805, but first successfully practised by the late Dr. J. Kearney Rodgers,[3] of New York. Dr. Henry J. Bigelow,[4] of Boston, in 1867, published an account of eleven cases, all but one successfully treated by this method, great care having been taken to preserve the periosteum. I have myself treated a considerable number of cases in a similar manner. When exercise is required in the open air, as when the patient's health has suffered from general debility, loss of blood, or protracted confinement, or where all

---

[1] Phila. Journ. Med. and Phys. Sci., vol. v. p. 116.
[2] Northwestern Med. and Surg. Journ., March, 1852, p. 409.
[3] N. Y. Med. and Phys. Journ., vol. vi. p. 521.
[4] Ununited Fractures, Boston, 1867.

hope of effecting a cure has been abandoned, great comfort will be experienced from the use of the excellent apparatus invented by Dr. Henry H. Smith,[1] of Philadelphia, and delineated in most of our treatises on surgery. In fracture of the tibia, attended with great overlapping or loss of substance, thereby rendering the limb comparatively useless, the fibula may occasionally be advantageously broken opposite to the original lesion, as suggested and successfully practised in several cases by Professor William H. Pancoast.

Children are liable to a form of injury of the osseous tissue, known as the green stick fracture, or as bending or incomplete fracture of the bones, the pieces most commonly affected being the ulna, radius, and clavicle. The humerus, femur, tibia, fibula, and ribs occasionally suffer in a similar manner, and even the bones of the head are not exempt from it. The first account of this singular lesion was from the pen of Professor Jurine, of Geneva, in 1810. The late Dr. John Rhea Barton gave a graphic description of it, illustrated by the narration of a number of cases, in the *American Medical Recorder* for 1821; and since that period the whole subject has received important additions from the labours of Professor Frank H. Hamilton, who has thoroughly investigated it in an elaborate series of carefully performed experiments upon the inferior animals, leaving nothing further to be desired upon this branch of surgical pathology.

The reduction of *Dislocations* has been greatly simplified during the last twenty years, chiefly through the genius and influence of American surgeons. The pulleys, formerly so much in vogue, and deemed, in many cases, indispensable to success, have become almost obsolete instruments; and, as to Jarvis's adjuster,[2] hardly any one thinks of employing it.

[1] Am. Journ. Med. Sci., Jan. 1848 and Jan. 1876.
[2] Lancet, 1846, vol. i.

These contrivances, thank fortune, have had their day. A new era has been inaugurated; science and common sense have taken the place of awkward and dangerous mechanical appliances, and the whole process has been rendered so simple that one is astonished that the revelation was so long in coming. Reduction by manipulation is now the order of the day, not only in simple cases, but even in the most complicated, as well as in many of the more protracted. Although this method is spoken of by Hippocrates, and although it occasionally succeeded in the hands of some of the practitioners during the last two hundred years or more, it remained for our countryman, Dr. William W. Reid,[1] of Rochester, New York, by a series of admirably conducted dissections, experiments, and observations, to generalize the method, and to establish the universality of its application. The paper in which the results of his labours are comprised, was published in 1855, and earned for him a wide-spread reputation. Like Byron, he woke up one morning and found himself famous; for he had justly earned his laurels. It does not in the least detract from the merits of Dr. Reid, when I add that he fell into error in referring the chief resistance to the reduction of dislocations to the action of the muscles at and around the injured joint. In many instances, indeed, as is proved when the muscular system is completely relaxed by anæsthesia, the obstacle is evidently caused by the resistance offered by the ligaments, especially in the ball-and-socket joints, as those of the hip and shoulder. In the former of these articulations, as was first shown by Professor Gunn,[2] of Chicago, and Professor Moore, of Rochester, and more recently by Professor Henry J. Bigelow,[3] of Harvard Uni-

---

[1] Buffalo Med. Journ., Aug. 1851, page 129, and N. Y. Journ. of Med., July, 1855, p. 55.

[2] New York Journ. of Med., Nov. 1853, p. 423.

[3] The Hip, Bigelow, Phila. 1869.

versity, the obstacle to restoration is largely, if not exclusively, due to the manner in which the head and neck of the thigh-bone are girt by the untorn portion of the capsular ligament. These facts are clearly stated by all these surgeons, and they have been made the subject of a beautiful and valuable monograph by the Boston surgeon, published in 1869. Dislocations of many months' duration have, in a number of instances, been more or less readily reduced by manipulation alone. In denouncing the pulleys as a relic of barbarous surgery, I do not mean to deny that they may not occasionally be employed with advantage; but their day, as a general principle, is certainly over, and we have no regret at parting with them.

The practice of reducing dislocations by simple manipulation is by no means a modern expedient. Distinct mention of it occurs in the writings of Hippocrates and Paul of Ægineta; it was successfully employed in the last century by Turner, Anderson, and other English surgeons; and in the early part of the present by Physick, of Philadelphia, and Nathan Smith, of New Haven. The late Mr. Morgan, of Guy's Hospital, London, was accustomed, it is said, for many years to avail himself exclusively of this method, asserting that the use of pulleys was wholly unnecessary. To Dr. Reid, however, is unquestionably due the credit of directing to it the attention of the profession in such a manner as to lead to its general adoption.

Every surgeon has occasionally experienced great difficulty in reducing dislocations of the thumb and fingers, especially of the former. Professor Dixi Crosby,[1] of New Hampshire, aware of this fact, adopted, in 1826, what was then a novel method, consisting simply in pushing the luxated phalanx forcibly back upon the metacarpal bone, until it forms a

---

[1] Am. Journ. Med. Sci., April, 1853, p. 401, and Boston Med. and Surg. Journ., Oct. 1, 1857, p. 172.

right angle with it, when, by strong pressure applied to its base from behind forward, it is readily carried by flexion into its natural position. Some European writers have claimed this operation for Mons. Gerdy, of Paris, but its original suggestion justly belongs to Dr. Crosby. The spatha of Dr. Richard J. Levis,[1] of this city, is a contrivance of great power, well adapted to the reduction of dislocations of the thumb and fingers, and is a valuable improvement upon the more ordinary procedures.

In dislocations of the sterno-clavicular and acromio-clavicular joints great difficulty, amounting occasionally to impossibility, is experienced in preserving the contact of the articular surfaces. To meet this contingency I suggested, many years ago, the importance of connecting the parts with strong silver wire; an idea first carried successfully into practice by the late Dr. Cooper,[2] of San Francisco, and soon afterwards by Dr. Hodgen, of St. Louis.

The subject of congenital dislocations of the hip-joint was ably illustrated by the late Dr. Carnochan. In an exhaustive memoir, published in 1850, he gave an able account of the lesion, accompanied by the narration of a number of cases and dissections.

Excessive suffering is occasionally experienced in old, irreducible luxations of the shoulder-joint from the pressure of the head of the bone upon the brachial plexus of nerves. In a case of this kind, in 1869, in a woman fifty years of age, under the care of Dr. Edward Warren,[3] formerly of Baltimore, prompt and permanent relief was afforded by the excision of the offending portion of bone.

To Professor L. A. Dugas,[4] of Georgia, is due the credit of having pointed out, as early as 1856, a most valuable

---

[1] Am. Journ. Med. Sci., Jan. 1857, p. 62.
[2] Ibid., April, 1861, p. 389.
[3] Baltimore Med. Journ., Sept. 1871, p. 532.
[4] Southern Med. and Surg. Journ., March, 1856, p. 131.

diagnostic sign of dislocation of the shoulder-joint. It is simply this, that, when the head of the humerus is thrown off from the glenoid cavity of the scapula, it is impossible for the patient, or the surgeon, to place the fingers of the injured limb upon the sound shoulder while the elbow touches the front of the chest.

The difficulty of effecting reduction in dislocations of the elbow backwards, in cases even of comparatively recent standing, is well known to surgeons. In several cases of this kind, in the hands of Dr. F. H. Hamilton[1] and Dr. Lewis A. Sayre, the object was readily attained by the subcutaneous division of the triceps muscle; and Dr. Waterman,[2] of Massachusetts, and myself have been equally successful by the method of forcible extension of the forearm.

In the cure of *Bony Anchylosis*, the world is indebted to American surgeons for several operations of an extremely ingenious character, since practised, more or less extensively, and more or less successfully in all parts of the world. Foremost among these operations is that of Dr. John Rhea Barton,[3] originally performed in 1826, by cutting out a V-shaped portion of the superior extremity of the femur in a sailor, aged twenty-one years, who had lost the use of his hip-joint from the effects of a fall on shipboard. Passive motion was instituted at the end of three weeks, and steadily maintained for four months, when the man had so far recovered as to be able to walk about with the aid merely of a cane. Eventually, however, the use of the artificial joint was completely lost In 1844, Dr. Gurdon Buck[4] modified the operation of Barton by attacking

---

[1] Hamilton, Fractures and Dislocations, 5th ed., p. 635, Phila. 1875.
[2] Bost. Med. and Surg. Journ., N. S. vol. iv., 1869.
[3] North Am. Med. and Surg. Journ., vol. iii. pp. 279, 400.
[4] Am. Journ. Med. Sci., Oct. 1845, p. 277.

the affected joint itself. The case was one of anchylosis of the knee-joint, from which, after cutting through the skin and muscles, and dissecting up the flap, he sawed out a V-shaped portion, extending nearly through the entire thickness of the femur, leaving indeed merely a little layer behind, which was then broken, when the limb was placed upon a double inclined plane, in an easy posture, at a suitable angle for future usefulness. An operation based essentially upon that of Barton, or, in other words, involving precisely the same principles, was suggested by Brainard,[1] of Chicago, and performed, in 1859, by Professor Joseph Pancoast,[2] the patient, a youth, making eventually, notwithstanding the formation of several abscesses and the occurrence of great constitutional trouble, a good recovery with a useful limb. The operation, performed for osseous anchylosis of the knee-joint, consisted in perforating the femur with a large gimlet, through a single opening in the skin, at half a dozen points, immediately above the articulation, and then forcibly breaking the bone. The procedure, it will be perceived, was a subcutaneous one. Dr. Brainard, in 1860, divided the femur through its condyles with a perforator; and, in 1861, I severed the connection between the articular extremities of the knee with the aid of a narrow chisel.[3] The operation thus performed is, I am warranted in declaring, perfectly free from danger, and should, in my opinion, founded upon the results of four cases, supersede every other devised for the purpose. Professor Sayre,[4] in 1862, in a case of bony anchylosis of the hip-joint, removed a segment of the femur above the small trochanter, and thus established a false joint, followed by a good use of the limb.

[1] Trans. Am. Med. Assoc., vol. vii. p. 557.
[2] Am. Journ. Med. Sci., April, 1868, p. 360.
[3] Ibid., April, 1868, p. 360.
[4] Trans. N. Y. State Med. Soc., 1863, p. 103.

America, if I mistake not, may claim priority in operations for the relief of anchylosis of the lower jaw. Dr. Carnochan, of New York, upwards of twenty years ago, in a case of this kind, in addition to the division of the masseter muscle, cut out a wedge-shaped portion of the body of the bone, according to Dr. Barton's principle, in order to form an artificial joint; the operation, however, proved to be a failure, and he, therefore, suggested, under similar circumstances, the removal of the entire half of the bone. In April, 1873, in a case of true anchylosis of the left temporomaxillary joint, in a girl seven years of age, I exsected the corresponding condyle of the bone, with the result of complete restoration of the movements of the jaw. The operation, in which I was kindly assisted by Drs. Levis, Barton, Hearn, and others, consisted in making a curvilinear incision in front of the ear, and after separating the condyle from its connections with the surrounding structures cutting it away with the pliers and chisel. Hardly any blood was lost; and, with the exception of a slight attack of erysipelas, the recovery was rapid and in every respect most satisfactory.

In the treatment of *Affections of the Joints*, American surgery stands pre-eminent. Physick, impressed with the great importance of complete and protracted rest in the management of this class of diseases, early in the present century, and long before the subject engaged the serious attention of our European brethren, availed himself of the employment of a splint, especially constructed for the purpose, to secure this end. He did not confine himself, as is generally supposed, in the use of this article, to the treatment of coxalgia, or scrofulous affections of the hip-joint, but insisted upon the indispensable necessity of rest in all maladies of the movable articulations, and also, not less emphatically, in the treatment of posterior curvature of the spine, known as Pott's disease. He was particularly successful in cases of coxalgia. The splint which he employed for securing complete rest to the affected structures was a curved one, constructed by Mr. Rush, of this city, a most skilful carver in wood; it extended from the middle of the side of the chest to within a short distance of the ankle, and was sufficiently wide to embrace nearly one-half of the parts to which it was applied. Instead of forcing the limb into a straight position, the splint was shaped to its angularities, and it was not until after the inflammation had been greatly reduced by the treatment that this was attempted, and that the original splint was replaced by a new one. The apparatus was carefully padded to prevent excoriation of the skin, and was confined to the trunk and limbs by appropriate bandages.

The period occupied in effecting a cure varied from six months to two years, the average being about twelve months. "During all this time," says Dr. J. Randolph, who published in the *American Journal of the Medical Sciences* for Feb. 1831, a full account of Dr. Physick's method of treating coxalgia, "the splint should be kept steadily applied; the surgeon, in fact, should not remove it until some time after all the symptoms and appearances of the disease have subsided." When this object has been attained, the exercise of the limb may be gradually and cautiously resumed. The treatment, adds Dr. Randolph, was particularly applicable to the earlier stages of the malady, prior to the occurrence of suppuration. Strict recumbency was enjoined for a long time, and a gentle laxative, consisting of jalap and cream of tartar, administered every few days. No material change was made in the diet, unless there was much inflammation, as evidenced by the hot, tender, and swollen condition of the parts, when—as, for example, in the case of a child six years of age—from four to six ounces of blood were usually taken by leeches as a preliminary measure. The drain from the bowels, established by the laxative, was considered as far more efficacious than the ordinary methods of counter-irritation. "Physick adapted the curved splint to the elbow, the knee, and the ankle, and frequently found this remedy to succeed when all others had failed." Unlike Sir B. C. Brodie and his followers, this great man looked upon ulceration of the articular cartilages, so common in this and similar diseases, as an invariable result of inflammation, and was thoroughly satisfied that rest was an indispensable element of treatment; an opinion now universally conceded by all enlightened practitioners.

It is needless to say that vast changes have taken place in the treatment of the diseases and injuries of the joints since the days of Physick and his earlier contemporaries; but the question of the great importance of rest, absolute and unconditional, stands precisely where it was placed soon after the commencement of the present century by the observations and teachings of the Father of American Surgery. With the improvements which have been made in the treatment of these affections in later years are honourably associated the names of Sayre, Davis, Taylor, and many other American practitioners.

To relieve the violent *inflammation* which occasionally occurs in the lower extremities, as in erysipelas, gunshot wounds, compound fractures, and compound dislocations, a very bold operation, consisting in the ligation of the femoral artery, was performed in 1813 by Dr. Henry M. Onderdonk,[1] of New York. The case was one of wound of the knee-joint, which had resisted all the usual means of treatment, but readily yielded to the remedy in question after grave fears had been entertained respecting the safety of the limb. In 1824 the operation was repeated by David L. Rogers,[2] also of New York, upon a man thirty years of

---

[1] Am. Med. and Phil. Reg., vol. iv. p. 176.
[2] N. Y. Med. and Phys. Journ., vol. iii. p. 453.

age, with results equally gratifying. From this time on nothing more was heard of this method of treatment until 1866, when it was revived, apparently without any knowledge that it had been done before, by Professor Henry F. Campbell, of Georgia, then a surgeon in the Confederate army. In an article in the *Southern Journal of the Medical Sciences*, he called the attention of the profession to the subject, and adduced a number of cases, chiefly of gunshot injuries of the bones and joints, in which it is stated to have yielded highly gratifying results. The only European surgeon, so far as I know, by whom ligation of the femoral artery has been practised for this purpose, is Mr. C. F. Maunder, of London, his case being one of violent inflammation of the limb, consequent upon a gunshot wound of the knee-joint. Of the value of this procedure nothing definite can, at present, be said, as the number of instances in which it has been employed is too limited to justify any positive opinion. One would suppose, judging from the results of general experience, that the usual depletory remedies, early and vigorously employed, and followed up by punctures, scarifications, and incisions of the affected structures, would, in almost every case, be sufficient to arrest any inflammation, however intense, without a resort to a measure, apparently, so fraught with danger as the ligation of the main artery of a limb.

The first application of the *trephine* for the relief of inflammation and abscess of bone, or inflammation of bone threatening to pass into necrosis, is generally ascribed to Sir Benjamin C. Brodie. The credit of priority, however, is justly due to Professor Nathan Smith,[1] of New Haven, who performed the operation as early as the latter part of the last century. In Nov. 1838 his son, the late Dr. T. Morven Smith, reported, in the *American Journal of the*

---

[1] Phila. Monthly Journ. of Med., June and July, 1827.

Medical Sciences, four cases, illustrative of the importance of the operation, in every one of which matter issued freely from the affected bone, although only a few days had elapsed since it was invaded by the disease. Of the nature of this mode of treatment in this class of affections it is impossible to form too high an estimate. Unfortunately it is seldom resorted to ; or, if employed, the operation is performed too late to be productive of much benefit. For some valuable remarks upon the pathology and treatment of this affection, with an account of the history of the operation, the reader is referred to an able article by the late Professor G. C. Blackman, in the *American Journal of the Medical Sciences*, for October, 1869.

Few subjects have been more closely or more thoroughly studied in this country than *Amputations*, and we accordingly find that the names of a considerable number of our surgeons are associated with "methods" of operating, either peculiar to themselves or modifications of the proceedings of others. Dr. John Warren,[1] a surgeon in the Revolutionary Army, and the first Professor of Anatomy and Surgery at Harvard University, was the first in this country, as far as my information extends, to remove the arm at the shoulder-joint, the operation having been performed in 1781. Amputation at this joint for gunshot injury was practised for the first time in 1813, by Dr. William Ingalls, of Boston. To Dr. Walter Brashear,[2] of Bardstown, Kentucky, belongs the honour of having led the way in amputations at the hip-joint. The case, which occurred in 1806, was a peculiar one. The thigh was at first removed in its continuity, but, as the bone was diseased in its entire length, it was disarticulated, and the patient, a lad seventeen years old, made an excellent recovery. The next case, also a successful one,

---

[1] Boston Med. and Surg. Journ., vol. xx. p. 210.
[2] Mott's Velpeau's Surgery, edited by Blackman, vol. ii. p. 270.

in the United States, was that of Dr. Valentine Mott,[1] in 1824. To these names may be added those of Buck, May, Bradbury, Van Buren, Joseph Pancoast, Paul F. Eve, S. D. Gross, J. Mason Warren, G. C. Blackman, J. H. Packard, Addinell Hewson, T. G. Morton, Whitcomb, Fauntleroy, William S. Forbes, D. H. Agnew, George A. Otis, and Frank F. Maury. In some of these cases the disarticulation was effected secondarily, some time after amputation of the thigh in its continuity. Pancoast and myself each had two primary cases, followed by recovery. The only successful example of primary amputation at the hip-joint, performed during the late war on account of gunshot injury, was that of Dr. Edward Shippen,[2] of Philadelphia. Dr. J. Mason Warren was the first Boston surgeon to remove the thigh at the hip-joint. Valuable statistics of this amputation have been published by Dr. Stephen Smith[3] and Dr. George C. Blackman.[4] The prevention of hemorrhage during this operation has always been a great desideratum with the surgeon, and happily, alike for science and humanity, this object was fully attained in Professor Pancoast's[5] first case, by the compression of the abdominal aorta by means of a modification of Signorini's tourniquet, an expedient the credit of which was unjustly claimed by our English cousins for Mr. Lister, of Edinburgh. The first operation in which this instrument was employed was performed in June, 1860, at the Pennsylvania Hospital.

Amputation at the knee-joint was originally performed in 1581, by Fabricius Hildanus, and in this country for the first time, in 1824, by Professor Nathan Smith,[6] of New

---

[1] Phila. Journ. Med. and Phys. Sci., vol. xiv. p. 101.
[2] Circular, No. 7, S. G. O., Washington, 1867.
[3] N. Y. Journ. of Med., Sept. 1852, p. 184.
[4] Western Lancet, vol. xvii., p. 7.
[5] Am. Journ. Med. Sci., July, 1866, p. 22.
[6] Am. Med. Rev., Dec. 1825, vol. ii. p. 370.

Haven, his patient making a prompt and thorough recovery. Velpeau made an attempt to revive the operation, and with this view, in 1830, published an elaborate paper, comprising an account of a number of successful cases in commendation of it. In this country the procedure has been warmly lauded by Dr Thomas Markoe,[1] of New York, and by Dr. John H. Brinton,[2] of Philadelphia, whose efforts have been greatly instrumental in promoting the generalization of the operation, by giving assurances, founded upon personal observations, of its comparative immunity from danger, and of its superiority over amputation of the thigh in its continuity. Of 164 cases of amputation at the knee-joint, tabulated in 1868 by Brinton, 53 were fatal, affording thus a mortality of about 32 per cent., or from one-fourth to one-sixth less than in amputation of the limb above the articulation. Of these cases 117 occurred in the practice of American surgeons The method of operation usually adopted in this country is that by anterior and posterior flaps. The employment of lateral flaps has been strongly recommended by Stephen Smith,[3] of New York, on the ground that the resulting stump is better adapted for drainage and less liable to injury from the pressure of the artificial limb. A case occasionally occurs, as in gunshot and other lesions, in which as many as three flaps may be required, as in several instances in the hands of Professor Pancoast.

Amputation of the ankle-joint was performed in this country for the first time in 1851 by myself with the aid of Professor Granville Sharp Pattison, before the medical class of the University of New York, during my connection with that school as Professor of Surgery, in the case of a young girl, the subject of caries of the tarsal and of some of the metatarsal bones. A good recovery followed; but,

[1] New York Journ. of Med., Jan. 1856, p. 9.
[2] Am. Journ. Med. Sci., April, 1868, p. 305.
[3] Ibid., Jan. 1870, p. 33.

owing to a return of disease, amputation of the leg was subsequently performed. The operation devised by Pirogoff, of Russia, a modification of that of Syme, has occasionally been successfully performed by our surgeons with the result of an excellent stump. Dr. Addinell Hewson,[1] of Philadelphia, has been particularly fortunate with his cases, several of which he has kindly afforded me an opportunity of inspecting. Chopart's amputation has also been repeatedly performed, and leaves in the main a better stump. In the cases in which I have had occasion to employ it, the result has been highly gratifying.

Amputation at the elbow-joint, now an accredited operation, was first performed on this side of the Atlantic in 1812 by Dr. Mann, U. S. A.,[2] during our late war with Great Britain. The operation leaves an excellent stump, and is less dangerous than amputation in the continuity of the arm, the same rule applying here as in amputation at the knee-joint in relation to the thigh.

Synchronous amputation is occasionally demanded in severe mutilation of the extremities consequent upon gunshot, railway, and other injuries. Examples of this kind, often followed by rapid recovery, have occurred within the last fifteen years in the practice of a number of our surgeons, as Carnochan, Eve, Seiler, S. W. Gross, Warren Stone, and John G. Koehler. The case of the latter is one of the most remarkable upon record, the operation involving the immediately successive removal of both legs and one arm, in a lad thirteen years of age, who made a speedy recovery.

The good luck which occasionally attends upon ordinary amputations in the hands of a skilful and judicious surgeon is well exemplified in the practice of Professor Eve. In a paper in the second volume of the new series of the *Southern*

---

[1] Am. Journ. Med. Sci., July, 1864, p. 121.
[2] Med. Repertory, 1822, N. S., vol. vii. p. 17.

Medical and Surgical Journal, this gentleman reports fourteen consecutive successful amputations of the thigh and leg, and fifty-one amputations in general without a single loss.

Amputation above the shoulder-joint, involving the removal of the superior extremity along with the clavicle and scapula, was first performed in this country, in 1836, by the late Dr. Dixi Crosby, Professor of Surgery at Dartmouth College, New Hampshire. The morbid growth, belonging to that class of tumours then and long afterwards denominated osteosarcomatous, measured thirty-seven inches in circumference at its widest part. Little blood was lost in the operation. The patient, a man thirty years of age, made a rapid recovery, but died twenty-eight months afterwards from an attack of paraplegia, due, as was conjectured, to malignant deposits in the lumber region of the spine. An abstract of this interesting case was communicated to me in 1870 by the distinguished operator, and a full account of it was published in the *New York Medical Record* for November, 1875, by his son Dr. A. B. Crosby, Professor of Anatomy at Bellevue Hospital Medical College.

An operation similar to that of Dr. Crosby was performed, in 1837, by Dr. Reuben D. Mussey, then of Hanover, New Hampshire, and afterwards of Cincinnati, Ohio; in 1838, by Dr. Amos Twitchell, of Keene, New Hampshire; in 1838, by Dr. George McClellan, of Philadelphia; and, in 1845, by Dr. David Gilbert, Professor of Surgery in the Medical Department of Pennsylvania College, Philadelphia. In Mussey's[1] case the removal of the scapula and clavicle was a secondary operation, performed five years after amputation of the arm at the shoulder-joint. In Gilbert's[2] case the body of the scapula and the greater portion of the collar-bone

---

[1] Am. Journ. Med. Sci., Feb. 1838, p. 390.
[2] Ibid., Oct. 1847, p. 360.

were retained. It is not positively certain that the entire clavicle was removed in the cases of Crosby and Twitchell; in that of McClellan a small portion was left, whereas in that of Mussey it was disarticulated at its junction with the sternum. The only instance of amputation above the shoulder-joint, antedating that of Professor Crosby, was one performed, in 1830, by Gaetani Bey, of Cairo, but this was a traumatic one, in which, after disarticulation at the shoulder-joint, the scapula was dissected out along with the projecting extremity of the clavicle; a very easy task as compared with the removal of these structures when involved in an enormous morbid growth. Dr. Gurdon Buck, of New York, in 1864, amputated the scapula and a part of the collar-bone in a case in which the arm had been removed by a previous operation. From this brief historical sketch of these exploits it is evident that Dr. Crosby, who was for many years Professor of Surgery at Dartmouth College, and for a long time one of the leading surgeons of New England, is justly entitled to priority in an operation which reflects so much credit upon our country.

In *Excision of the Bones and Joints* no country has a better record than ours. Indeed, it is in this field that many of our surgeons have achieved some of their most brilliant triumphs. Commencing with the extirpation of the clavicle, we find that this bone was removed by Dr. Mott,[1] in 1828, in its entirety, on account of a sarcomatous enlargement, measuring four inches in diameter at its base. The operation lasted nearly four hours, and was one of the greatest possible delicacy, requiring, as is apparent from the size and situation of the tumour, an extraordinary amount of skill for its successful execution. Upwards of forty vessels were secured. The patient made an excellent recovery, and with the aid of an apparatus constructed for the purpose obtained

---

[1] Am. Journ. Med. Sci., Nov. 1828, p. 100.

a good use of his arm. Excision of this bone for similar reasons has been executed by John C. Warren, Cooper, Curtis, Eve, Palmer, and others. Dr. Charles McCreary,[1] of Kentucky, in 1813, excised the clavicle at its articulations for scrofulous caries; and similar operations have been performed by Wedderburn, Blackman, Fuqua, and other surgeons. As a brilliant and daring exploit, the case of Dr. Mott is without a parallel in the annals of surgery.

The scapula has been repeatedly removed, by the surgeons of this country, generally on account of sarcomatous or enchondromatous disease; the principal operators having been Mussey, of Cincinnati, Gross, of Philadelphia, Hammer, of St. Louis, Rogers, of New York, and Schuppert, of New Orleans. In the cases of Hammer and of Rogers,[2] the excision was effected at two periods; in the first after the lapse of a few days, in the second of a few months.

With excision of portions of the ribs of greater or less length, are associated the names of a number of American surgeons, among whom may be especially mentioned John C. Warren, George McClellan, William Gibson, and William A. McDowell.[3] In the latter case the sixth and seventh ribs were disarticulated at their connection with the vertebræ; the only one of the kind, I believe, on record. I have myself done a good deal of work in this direction. The sixth volume of the *Philadelphia Journal of the Medical and Physical Sciences* records a remarkable case, in which Dr. Milton Antony, of Georgia, alleges to have removed the fifth and sixth ribs, which were extensively carious, along with two-thirds of one of the lobes of the right lung; the patient surviving the operation nearly four months.

Excision of the coccyx, performed as early as 1832, by

---

[1] Gross' History of Kentucky Surgery, p. 180.
[2] Am. Journ. Med. Sci., Oct. 1868, p. 359.
Am. Med. Recorder, vol. xiii., 1828, p. 113.

Dr. Josiah C. Nott,[1] of Mobile, Alabama, on account of a neuralgic affection now known as coccyodynia, was a pioneer operation, highly creditable to the originator. The bone, hollowed out into a mere shell, and the seat of severe and intractable suffering, was disarticulated at the second joint and carefully detached from its ligamentous and muscular connections. Although the wound was slow in healing, the patient, a lady, twenty-five years old, eventually completely recovered. Since that time the operation has been repeated in a number of instances, always with excellent results: a full account of it was published by the late Sir James Y. Simpson, of Edinburgh, to whom has occasionally been erroneously awarded the credit of having been the first to suggest and to perform it.

The only cases of excision of the head of the humerus claiming attention in a paper like this, are those reported by Professor Warren (*vide* p. 154), formerly of Baltimore, and lately in the service of the Khedive of Egypt, and of the late Professor Blackman, of Cincinnati, in the former of which this bone was removed, in 1868, on account of the pressure which it produced upon the axillary plexus of nerves, caused by downward displacement; and in the latter, in which, in 1870, it was exsected for a similar reason on account of arthritic enlargement. Both operations were successful in relieving the excruciating pain, for the relief of which they were performed. Practically these two cases present great points of interest as examples for imitation in similar lesions. Excision of the elbow-joint, originally executed by the elder Moreau, in 1794, was first performed in this country by Dr. John C. Warren,[2] in 1834. Of excision of the wrist-joint I am unable to say who led the way. Dr. Robert B. Butt,[3] of Virginia, exsected the lower two-thirds

---

[1] Am. Journ. Med. Sci., Oct. 1844, p. 544.
[2] Hodges, Excision of Joints, p. 69.
[3] Phila. Journ. Med. and Phys. Sci., vol. x. p. 115.

of the ulna in 1825; the entire bone was removed by Dr. Carnochan,[1] of New York, in 1853; by Dr. C. T. Muscroft,[2] of Cincinnati, in 1870; and by Dr. Joseph C. Hutchison,[3] of Brooklyn, in 1873. The entire radius was exsected by Carnochan[4] in 1854; and a few years afterwards a similar operation was performed by Professor Choppin, of New Orleans. Both these bones, with the exception of the inferior extremity of the former, were removed, in 1853, by Dr. Compton,[5] of New Orleans. All these cases made good recoveries with useful limbs. The olecranon process was exsected by Dr. Gurdon Buck[6] in 1842, on account of chronic hypertrophy of its substance interfering with the functions of the elbow-joint.

Excision of the hip-joint, originated by Mr. Anthony White, of London, in 1822, was first performed in the United States by Professor Henry J. Bigelow,[7] of Boston, in 1852. Since then it has become a very common procedure, especially for the cure of coxalgia, Professor Sayre, of New York, alone having executed it fifty-nine times, with a degree of success reflecting the highest credit upon his judgment as a practitioner, and upon his dexterity as an operator. Of these cases thirty-nine were alive at the close of 1875, with a good use of the corresponding limb. Excision of the great trochanter was first performed among us by Professor Willard Parker. Excision of the knee and ankle-joints has not been often practised by our surgeons, and I am unable to state who is entitled to priority. A case of resection of the entire fibula for fibro-cartilaginous degenera-

---

[1] Am. Med. Monthly, March, 1854.
[2] Cin. Lancet and Obs., Aug. 1870, p. 449.
[3] Am. Journ. Med. Sci., Jan. 1874, p. 96.
[4] Ibid., April, 1858, p. 363.
[5] Mott and Blackman's Velpeau, vol. ii. p. 460.
[6] Am. Journ. Med Sci., April, 1843, p. 297.
[7] Ibid., July, 1852, p. 90.

tion of that bone was reported, in 1858, by Dr. A. R. Jackson,[1] now of Chicago. The calcaneum has been removed by Carnochan, Morrogh, Greenleaf, McGuire, and others; the astragalus, by Peace; the calcaneum and astragalus, by T. G. Morton; the calcaneum and cuboid bone, by J. T. Bradford, of Kentucky. Professor H. J. Bigelow, in 1855, cut away all the tarsal bones, excepting the astragalus and calcaneum, together with the head of the second and third metatarsal bones; and, in 1874, Professor P. S. Conner,[2] of Cincinnati, resected the metatarsus, anterior tarsus, and portions of the astragalus and calcaneum, followed by recovery with a useful foot.

Trephining of the skull for the cure of epilepsy was originally performed by La Motte, and revived by Cline, of London, in the early part of the present century. In this country the operation seems to have been first performed by Professor Dudley, of Lexington, Kentucky, who, in 1828, published an interesting article upon the subject in the first volume of the *Transylvania Journal of Medicine*, in which he detailed the particulars of five cases, of which three were entirely relieved, the first having occurred in 1819. In the eleventh volume of the *American Journal of the Medical Sciences*, the same surgeon has a paper on the "Use of the Trephine in Epilepsy," being, as he states, his sixth successful case. Dr. John G. F. Holston, of Ohio, has reported seven cases, in one of which a cure was effected after twenty years of suffering. Dr. J. T. Gilmore, of Mobile, has had five cases, with three cures and two deaths, the latter having been attended with the loss of some cerebral substance. A number of examples have been reported in which our surgeons trephined the skull and tied the carotid artery for the cure of this disease. In some of these cases

---

[1] Am. Journ. Med. Sci., Oct. 1858, p. 357.
[2] Ibid., July, 1875, p. 86.

both the common carotids were secured after a variable interval. Of the utility of trephining in certain forms of epilepsy there can be no doubt; but that the operation, however well performed, is generally one of great danger, is equally true. Of four cases in my hands the result in three was fatal. For valuable statistical facts relative to this operation the reader is referred to instructive papers, by Dr. Stephen Smith,[1] of New York, and Dr. J. S. Billings, U. S. A., in the *Cincinnati Lancet* for June, 1861.[2]

Puncture of the head for the cure of hydrocephalus has been repeatedly performed by American surgeons, Physick[3] being one of the first, if not the first, to undertake it on this side of the Atlantic, his case having occurred in 1801. Dr. L. A. Dugas,[4] of Georgia, has reported a case in which he tapped a child's head seven times, and evacuated sixty-three ounces of water, the patient surviving nearly four months. Physick was of opinion that the operation would be more likely to terminate favourably if it was performed without the aid of compression, the puncture being made very small, so as to allow the water to drain off very gradually.

Dr. Amasa Trowbridge,[5] of Watertown, New York, in 1829 reported three cases of hydrorachitis successfully treated, one with the knife, and the others with the wire ligature. Mott, Nott, Sayre, and several other American surgeons, have been equally fortunate with excision. In an instance in the hands of Dr. Charles Skinner,[6] of North Carolina, a tumour of this kind was punctured seventy times without any bad effects. The treatment of the late Dr.

---

[1] New York Journ. of Med., March, 1852, p. 230.
[2] See Am. Journ. Med. Sci., July, 1861, p. 299.
[3] Phila. Journ. Med. and Phys. Sci., vol. xiii. p. 316.
[4] Am. Journ. Med. Sci., Aug. 1837, p. 536.
[5] Boston Med. and Surg. Journ., vol. i. p. 753.
[6] Am. Journ. Med. Sci., Nov. 1836, p. 109.

Brainard,[1] of Chicago, for the cure of this affection, is well known; he published a number of successful cases, but, it must be confessed, the remedy in the practice of others has been, for the most part, a signal failure.

Trephining of the vertebræ in cases of fracture and dislocation has been repeatedly performed, but in no instance with any permanent benefit. One of the first of these operations in America was that of Dr. John Rhea Barton, soon followed by those of Alban Goldsmith and J. K. Rodgers, and, more recently, by those of Blackman, Potter, Hutchison, Stephen Smith, and others.

America may justly claim the honour of having led the way in extirpations of the *upper jaw*. Small portions, it is true, had been chipped off in the eighteenth and even in the seventeenth century; but the first grand and difficult operation of the kind of which we have any knowledge, was performed in 1820, by Dr. Horatio G. Jameson,[2] of Baltimore, who took away nearly the entire bone on one side, the roof of the antrum alone being left, as it was not involved in disease. Resection of both bones, a still greater triumph of surgery, was first performed in 1824, by David L. Rogers,[3] of New York, who carried his incisions as far back as the anterior limits of the pterygoid processes of the sphenoid bone, his patient, like that of Jameson, also making a good recovery. Since the feasibility of these procedures was thus established, excision of the upper jaw has been performed in many hundred cases both in this country and in Europe, and long ago took its place among the approved operations in surgery. I have myself on several occasions removed the entire upper jaw along with considerable portions of the palate and sphenoid bones; and I have been present on

---

[1] Am. Journ. Med. Sci., July, 1848, p. 262.
[2] Am. Med. Rec., vol. iv. p. 222.
[3] New York Med. and Phys. Journ., vol. iii. p. 301.

several occasions in which Professor Joseph Pancoast performed a similar operation. Very considerable portions of the upper jawbone have been removed by some of our surgeons without any external incision in the cheek, Dr. William E. Horner[1] having, I believe, been the first to make such an effort.

In regard to excision of the lower jaw, the first case upon record is that of Dr. Deaderick,[2] of Tennessee, who, in 1810, removed a section of the bone extending from near its angle to the centre of the chin, on account of a morbid growth, now generally supposed to have been an enchondroma. The result was a complete success, the patient surviving the effects of the operation many years. Unfortunately for the reputation of Dr. Deaderick, no account of this operation was published until 1823; a circumstance which for a time enabled other surgeons to bring forward their supposed claims to priority. The honour of performing the more bold and formidable operation of disarticulating this bone, and sawing it off at the chin, belongs to Valentine Mott,[3] who achieved it in 1821. In 1823, Dr. George McClellan[4] removed all that portion of the bone immediately anterior to its two angles. In the *New York Journal of Medicine* for 1850, a case is reported in which Professor Ackley, of Cleveland, Ohio, is said to have successfully exsected the entire bone on account of osteosarcoma, the patient being in good health two years after the operation. In the *American Journal of the Medical Sciences* for 1831, Dr. John Rhea Barton has described a case in which, on account of a so-called epulis, he exsected a longitudinal section of the lower jaw, leaving the base of the bone as a means of pre-

---

[1] Med. Examiner, Jan. 1850, p. 16.
[2] Am. Med. Recorder, vol. vi. p. 516.
[3] New York Med. and Phys. Journ., vol. i. p. 385.
[4] McClellan's Surgery, p. 365.

serving the countour of the face. Dr. James R. Wood,[1] of New York, in 1856, succesfully removed the entire bone on account of phosphorus disease.

For the relief of false anchylosis of the lower jaw depending upon permanent contraction of the masseter muscle, caused by inordinate ptyalism, at one time so common in this country, subcutaneous division of the muscles in question was formerly practised; the operations for this purpose having been inaugurated almost simultaneously by J. W. Schmidt,[2] of New Orleans, and Mott and Carnochan,[3] of New York. To facilitate the restoration of the joint, and prevent recurrence of the contraction, Mott,[4] Barton, and other American surgeons employed forcible dilatation by means of screws and levers, ingenious modifications of the instrument originally devised by Scultetus.

The difficult operation of extirpating the *parotid gland*, long considered as impracticable, has been repeatedly performed in the United States, the first instance having occurred in the hands of Dr. John Warren,[5] of Boston, in 1804. The late Dr. Geo. McClellan,[6] of Philadelphia, performed the operation altogether nearly a dozen times, and the brilliant results attending his cases exerted a powerful influence in generalizing the procedure by inspiring surgeons with confidence in its feasibility. Doubtless some of the cases that have been reported as excisions of the gland were cases simply of morbid growths developed upon its surface or embedded in its substance; but that most of them were genuine examples of extirpation is sufficiently proved by the high character of the respective operators. The names of

[1] New York Journ. of Med., May, 1856, p. 301.
[2] Am. Journ. Med. Sci., Oct. 1842, p. 516.
[3] Mott's Velpeau by Townsend, vol. ii. p. 20, appendix.
[4] Am. Journ. Med. Sci., Nov. 1829, p. 102.
[5] Surgical Observations on Tumours. Warren, p. 287.
[6] McClellan's Surgery, p. 333.

such men as Valentine Mott, George Bushe, N. R. Smith, William E. Horner, J. Mason Warren, Joseph Pancoast, H. H. Toland, and Frank H. Hamilton, not to cite others of great respectability, is a sufficient guarantee that they are capable of forming a correct estimate of what they see and do ; in other words, know what they are about. The day for denying the possibility of such an operation is long since passed. The procedure, however, is certainly one of the most difficult in surgery.

Partial excision of the *tonsils* has superseded the removal of these organs with the double canula, practised fifty years ago. Most of our surgeons, if I mistake not, perform the operation with the volsella and probe-pointed bistoury, but there are some who still prefer the ingenious tonsillotome originally devised by Physick[1] in 1828, and modified by Gibson, Hosack, Fahnestock, Rogers, Cox, Mitchell, and other practitioners. When the enlarged tonsil is very brittle and unusually vascular, a condition not unfrequently followed by more or less copious hemorrhage, removal should, if possible, be effected by crushing with the tonsil écraseur, or chain instrument, suggested for this purpose by Dr. S. W. Gross, of Philadelphia. To Dr. Alexander E. Hosack, of New York, is usually ascribed the credit of having first called the attention of the profession to excision of the tonsils with the knife, his paper upon the subject having been published in the *American Journal of the Medical Sciences* for Feb. 1828.

In carcinomatous, sarcomatous, and other tumours of the tonsils, extirpation of the diseased mass, when this is deemed proper, is usually performed through the mouth, notwithstanding its well known difficulties, large portions of the morbid growth being often left behind in consequence. In a case of encephaloid of the tonsils under the care, a few years ago, of Dr. Cheever,[2] that gentleman attained his

[1] Am. Journ. Med. Sci., Feb. 1828, p. 116.
[2] First Med. and Surg. Reports, Boston City Hosp., p. 390.

object by working his way through the upper and lateral parts of the neck. The operation, the first of the kind, if I mistake not, upon record, and one which reflects great credit upon the skill and intrepidity of the Boston surgeon, necessitated the division of the stylohyoid and styloglossal muscles, together with the separation of the fibres of the superior constrictor of the pharynx, through the interspace of which the diseased mass was approached and removed. Twelve ligatures were applied. The wound was completely closed at the end of a month, unpreceded by any untoward occurrence.

Numerous ingenious instruments for the extraction of foreign bodies from the œsophagus have been devised by American surgeons. The forceps invented by the late Dr. Thomas Bond,[1] of this city, are widely known, and have served as models for the various modifications now before the profession in this and other countries. Dr. Physick,[2] early in his professional life, constructed a stomach tube for removing poisons from the stomach without being aware he had been anticipated by Dr. Monro, of Edinburgh.

Œsophagotomy has been repeatedly performed. For the relief of organic stricture of the œsophagus it was, if I mistake not, first performed in this country by the late Dr. John Watson, of New York. The case is reported at length in the *American Journal of the Medical Sciences* for 1844. Dr. David W. Cheever,[3] of Boston, has performed the operation three times successfully for the removal of foreign bodies in this passage; and in 1867 it was performed with equally fortunate results by Dr. Alfred Hitchcock,[4] of Fitchburg, Massachusetts.

For the removal of fibroid tumours of the *nose*, very

---

[1] N. A. Med. and Surg. Journ., vol. vi. p. 278.
[2] Am. Med. Recorder, vol. x. p. 322.
[3] Œsophagotomy, Cheever, Boston, 1868.
[4] Boston Med. and Surg. Journ., July 16, 1868.

severe, difficult, and bloody operations are not unfrequently necessary. In a case of this kind, in which an immense fibroid growth filled the entire nostril and dipped far down into the pharynx, Dr. Mott,[1] in 1841, succeeded in effecting complete riddance by approaching the morbid mass through an opening in front of the face made by dividing the nasal and maxillary bones The operation, one of the most extensive and intricate of the kind over performed, and the first of the kind ever done in this country, was followed by the most gratifying results. A case in which a large fibroid polyp was successfully removed from the base of the cranium through the hard and soft palate by Professor Paul F. Eve, is recorded in the *Southern Medical and Surgical Journal* for 1836. The patient was alive and doing well twenty years after the operation.

It is a singular fact that men, living in different countries, without intercommunication, occasionally effect simultaneous, or coincident discoveries or inventions. This seems to have been the case in regard to *staphylorraphy* or palate suture, first methodized by Roux, of Paris, in 1819, and performed the following year by Dr. John C. Warren,[2] of Boston, without any knowledge that it had previously been attempted in Europe. Such an occurrence can readily be believed when it is recollected that at the time adverted to, the intercourse between this country and Europe was comparatively infrequent, and knowledge traversed the Atlantic Ocean only at long intervals. However this may be, Dr. Warren deserves great credit for the simplicity and efficiency of his operation. The second case of staphylorraphy is said to have occurred in the practice of Dr. George McClellan in 1823, and the third in that of Dr. Alexander H. Stevens, of New York, in 1826. Since that time it has

---

[1] Am. Journ. Med. Sci., Jan. 1842, p. 257, and Jan. 1843, p. 87.
[2] Ibid., Nov. 1828, p. 1.

been performed by most of our distinguished practitioners, but by none so frequently as by the late Dr. J. Mason Warren, of Boston, who, up to 1860, a few years prior to his death, had had 100 cases. It has been generally believed that Sir William Ferguson was the first to call attention to the importance of dividing the muscles in the arches of the palate as a means of facilitating the reunion of the edges of the fissure, but it is now well ascertained that this credit belongs to Dr. J. Mason Warren, who published an account of the procedure in the *New England Quarterly Journal of Medicine and Surgery* for April, 1843, thus antedating the paper of the British surgeon by at least eighteen months. Be this as it may, the importance of this suggestion has, judging from my own experience, been greatly exaggerated; for it is only in cases attended with excessive width of gap that the measure is at all necessary, either as it respects the promotion of reunion of the borders of the opening or the subsequent improvement of speech. Numerous instruments, some of them of a very ingenious and useful character, have been devised for facilitating the different steps of the operation, among others by Dr. William Gibson, of Philadelphia, Dr. A. E. Hosack, and Dr. George Bushe, of New York, and Dr. Mettauer, of Virginia.

Dr. William S. Forbes recently called attention to a new operation devised by him for the cure of " Certain Cases of Cleft Palate" attended with fissure of the uvula. His account of the procedure, illustrated by drawings and two successful cases, will be found in the *Transactions of the College of Physicians of Philadelphia* for 1875.

In a paper in the *St. Louis Medical and Surgical Journal* for January, 1875, Dr. Prine, of Jacksonville, Illinois, has thrown out certain suggestions designed to facilitate the performance of staphylorraphy and to insure its more perfect success. The three new points which he claims are, the employment of the galvano-cautery, the introduction of what he styles automatic needles, and a new interpretation of the functions of the muscles in the pillars of the fauces, the division of which he declares to be not only useless but positively injurious.

Palatoplasty, uranoplasty, or staphyloplasty is an operation designed to close any opening, congenital or accidental, that may exist in the horizontal plates of the palate and maxillary bones. The operation was originally suggested and first practised by J. Mason Warren, who published an account of it in the *New England Quarterly Journal of Medicine and Surgery* for 1843. Of his success a good idea may be formed when it is stated that of twenty-four cases only one failed. Speech, deglutition, and local comfort were more or less improved in nearly every instance, although in several the loss of substance was unusually great.

Amputation of the *tongue* for the relief of congenital hypertrophy of that organ was performed for the first time, in the United States, by Dr. Thomas Harris,[1] of Philadelphia. Five years afterwards he repeated the operation, which, since that period, has been executed by different surgeons, most of the cases having made a good recovery. In an instance recently in charge of Professor N. R. Smith,[2] of Baltimore, the operation was successfully performed with a stout silk ligature.

Extirpation of the tongue for the relief of malignant disease has been practised, with more or less success by a number of American surgeons, among others by Dr. J. T. Gilmore, of Mobile, who has reported the particulars of four cases.

For the cure of salivary fistule, whether the result of injury or of ptyalism, an ingenious operation was devised by the late Professor W. E. Horner,[3] of Philadelphia, consisting

---

[1] Am. Journ. Med. Sci., Nov. 1830, p. 17.
[2] Ibid., April, 1875, p. 429.
[3] Smith's Surgery, Phila. 1863, vol. ii. p. 161.

in making an opening with a sharp punch while the cheek is firmly pressed against a strong spatula held on the inside of the mouth, and then uniting the edges of the wound in the skin with several points of twisted suture, thus compelling the salivary fluid to resume its natural direction.

Cheiloplasty and genioplasty, operations performed for the restoration of the lips and cheeks, are greatly indebted for many of their most successful applications to the genius and labours of Mott, Pancoast, Mütter, Hamilton, Buck, and others. The most remarkable results have followed the efforts of some of these distinguished surgeons in cases, apparently, at first sight, of the most unpromising character.

Extirpation of the *thyroid gland* on account of hypertrophy, cystic enlargement, or malignant disease, is very justly regarded as one of the most difficult and dangerous of operations; difficult because of the delicacy and importance of the structures involved, and dangerous because, first, of the excessive bleeding so liable to attend it, however carefully executed, and, secondly, the intensity of the resulting inflammation of the air-passages. For these reasons it is seldom undertaken. The first case of this kind in this country occurred, in 1807, in the practice of Dr. Charles Harris, of New York. The tumour, separated by enucleation, was a large one, and, although the deep-seated parts of the neck were extensively denuded, the patient rapidly recovered without any bad symptoms. The late Professor Blackman, of Cincinnati, reported one successful case; and in 1871 Dr. W. Warren Greene,[1] of Portland, published an account of three examples in which he was equally fortunate. In two of these the tumours were of great size, the weight of one being estimated at five pounds. From a letter recently received from Dr. Greene, I learn that he has extirpated the thyroid gland altogether in seven instances, with five

---

[1] Am. Journ. Med. Sci., Jan. 1871, p. 80.

recoveries and two deaths. Dr. F. F. Maury,[1] of Philadelphia, has had two cases of extirpation of goiterous growths, one followed by recovery, the other by death.

For the relief of hydropericardium *paracentesis* is occasionally required, and, although the operation has been repeatedly performed, both in this country and in Europe, it is believed that the first successful case was one which occurred in the hands of Dr. John C. Warren,[2] of Boston. Velpeau, who was thoroughly acquainted with the history of the operation, in his great work on surgery, expresses the opinion that it was, at the time at which he wrote, the only instance of the kind worthy of entire credence. The operation was performed with a small trocar and canula through the sixth intercostal space, the quantity of serum evacuated being between five and six ounces.

Thoracentesis for the removal of accumulations of serum or sero-purulent fluid in the pleural cavity deserves passing notice here on account of the great success which has attended its performance in the practice of that distinguished physician, Dr. Henry I. Bowditch, of Boston, who was the first on this side of the Atlantic to illustrate the great value of the operation. Indeed, it is not too much to say that he has done more by his labours and writings to diffuse a knowledge of the importance of thoracentesis than all other practitioners together. In a letter with which Dr. Bowditch recently favoured me, dated November 30th, 1875, he informs me that he had performed the operation 325 times, the number of his patients being 205. Dr. Bowditch is evidently master of the field in Massachusetts; indeed, I am not aware that any other practitioner, either in America or in Europe, can show such vast statistics. Of his success I am unable to speak, but it has been highly flattering, and

---

[1] Am. Journ. Med. Sci., Jan. 1873, p. 281.
[2] Smith's Surgery, Phila. 1863, vol. ii. p. 358.

his regret in common with most physicians and surgeons is that the operation is too often employed as a dernier ressort, when, as most generally happens, there is no chance whatever of permanent relief. The instrument used by Dr. Bowditch[1] is a small exploring trocar, to which, at the suggestion of Dr. Morrill Wyman, of Cambridge, he attaches, when necessary, a suction apparatus, similar to the ordinary stomach pump, thus literally anticipating Dieulafoy by nearly a quarter of a century in the employment of what is now known as the aspirator. In speaking of the operation, Dr. Bowditch remarks: "It was aspiration, pure and undefiled, and no mere assertion can make it different." In the thoracentesis of former times a much larger instrument was used, and the surgeon generally waited for "pointing." In the modern operation, which has been proved to be as innocuous as it is simple and satisfactory, the fluid is drawn off as early as possible. The instrument is easily introduced, without any liability of admitting air, or of leaving any fistulous opening. The place selected for the puncture, when practicable, is the back, in the eighth or ninth intercostal space, below the scapula, on a line with its inferior angle.

To the late Dr. Horace Green, of New York, the profession is indebted for having paved the way in this country in the treatment of the surgical affections of the *larynx*.[2] He was the first among us to demonstrate the possibility of applying nitrate of silver to this tube for the cure of inflammation, and considering that this was done years before the invention of the laryngoscope, he deserves not a little credit for his efforts to popularize this mode of medication, the value of which is now universally acknowledged. To Dr. Green also, to Dr. Gurdon Buck, and to Dr. Willard Parker, great credit is due for their contributions to the advance-

---

[1] Am. Journ. Med. Sci., Jan. 1852, p. 103, and April, 1852, p. 320.
[2] *Vide* Am. Med. Monthly, April, 1854, p. 241.

ment of our knowledge of morbid growths of the larynx, and what were considered at the time the best means of removing them. The instrument of Dr. Buck[1] for incising the lips of the mouth of the larynx in œdema of the glottis is well known.

The treatment of Wounds of the *Intestines* has long been an object of interest with practitioners. In this country the first attempt to place it in its true light was made by Dr. Thomas Smith, in his Inaugural Essay on "Wounds of the Intestines," presented for the degree of Doctor of Medicine in the University of Pennsylvania in 1805. This was followed, in 1812, by the treatise of Mr. Benjamin Travers, of London; and, in 1843, by my *Experimental and Critical Inquiry into the Nature and Treatment of Wounds of the Intestines*. This work was founded upon nearly seventy experiments upon dogs, instituted with a view of determining, if possible, the best mode of management in this particular class of lesions. Without going into details, it will be sufficient to state, as the legitimate conclusions from these researches, first, that all wounds of the bowels, however small, and of whatever direction, are almost inevitably followed by fecal effusion, unless sewed up before the parts are restored to the abdominal cavity; and, secondly, that the simple interrupted suture, made with a delicate, well-waxed silk or flax ligature, placed at a distance of a line and a half apart, and tied into a double knot, with the ends cut off close, answers, as a general rule, far better than any of the more complicated plans described in our works on surgery. These facts are now so well proved that they should be accepted as established principles of practice. As another deduction from these experiments, I ascertained that all penetrating wounds of the abdomen must be treated with the deep suture, passing close down

---

[1] Am. Journ. Med. Sci., Jan. 1840, p. 249.

to the peritoneum, otherwise, when the patient recovers, hernia will be inevitable. This circumstance is now well known, but it was not known a third of a century ago, when these experiments were performed.

In connection with this subject it may be stated that, in several cases of strangulated hernia, attended with gangrene of the intestine, occurring in the practice of American surgeons, the affected portion of the tube was excised, and the ends stitched together with the interrupted suture, followed by rapid recovery. In one instance in the hands of the late Dr. Charles A. Luzenberg,[1] of New Orleans, six inches of gut were thus successfully excised, the ligatures coming away at the end of thirty-five days. In a still more remarkable case, Dr. A. Brigham,[2] of Utica, removed seventeen inches of the small intestine, and yet the patient made an excellent recovery.

The radical cure of hernia has engaged much attention in this country, and some very ingenious modifications of different methods of treatment have been suggested. Among our countrymen who have been particularly conspicuous in interesting themselves in operative procedures of this kind, the names of Jameson, Watson, Pancoast, Riggs, Armsby, and Agnew deserve special mention.

The subcutaneous division of the stricture in strangulation in cases of old hernia, a hazardous operation in the hands of ordinary surgeons, has been practised successfully by the Professors Pancoast, father and son, of Philadelphia. I am not aware that the operation has ever been performed by any other practitioners in this country.

The improvement of trusses for the retention of the bowel in hernia and for the radical cure of this affection has long been an object of deep study on the part of our surgeons, as

[1] Gross' Am. Med. Biography, p. 556.
[2] Am. Journ. Med. Sci., April, 1845, p. 355.

is proved by the numerous inventions before the profession, not a few of which display uncommon ingenuity in the adaptation of correct principles and in the accuracy of their construction. Among the very best of these contrivances, if indeed not the best, are the trusses devised by the late Dr. Heber Chase, of Philadelphia, instruments which combine, in an eminent degree, all the requisite qualities of such supports. The ingenious inventor, a member of the profession, devoted many of the best years of his life to their improvement, and the result is that, although he long since passed away, they still command the confidence of our most enlightened and experienced practitioners.

To Physick[1] we are indebted for the first accurate account of an affection of the lower bowel, known as the cystic, sacculated, or sacciform rectum. It was originally described by him in the early part of the present century, and, as far as my knowledge extends, no mention of it is made in any English work, although it is of sufficiently frequent occurrence, especially in advanced life. Physick not only pointed out its pathology and mode of formation, but its proper treatment. Until the early part of the present century surgeons had no correct or fixed ideas respecting the situation of the internal aperture in anal fistule, and the consequence was that many severe and even dangerous operations were performed that might otherwise have been avoided. Professor Ribes, of Paris, led the way in the investigations which finally cleared up this matter; a path soon afterwards successfully explored by the late Professor Horner,[2] of Philadelphia, who by a series of careful dissections, fully verified the observations of the French anatomist, and thus definitively settled a question which had so long agitated the surgical profession in different parts of the world.

[1] Am. Cyclop. of Prac. Med. and Surgery, vol. ii. p. 123.
[2] Ibid., p. 82.

For the cure of anal fistule, I have for the last thirty years performed a very simple operation, consisting in the introduction of a flexible grooved director through the track into the bowel, drawing out the extremity of the instrument with the index finger previously inserted into the tube, and then, while the point of the instrument rests upon the opposite nates, dividing the overlying structures with a narrow bladed bistoury. I mention this operation because in Europe, and indeed, I believe, also, in this country, the old method of cutting from within outwards with the end of the finger hooked over the extremity of a probe-pointed bistoury in the rectum is the one still almost universally pursued.

Gastrotomy has been performed several times by American surgeons for the removal of foreign bodies from the stomach. Among other cases is the remarkable one of Dr. John Bell,[1] of Wapello, Iowa, in which a bar of lead, ten inches in length by upwards of six lines in diameter, weighing one pound, was successfully removed. Gastrotomy for the relief of suffering caused by organic stricture of the œsophagus has, I believe, been performed but once in this country, namely, in 1869, the operator being Dr. Frank F. Maury,[2] of Philadelphia. Death ensued at the end of seventeen hours. Dr. Samuel White,[3] of Hudson, New York, early in the present century, performed enterotomy successfully for the removal of a large teaspoon, swallowed in a paroxysm of delirium.

Extirpation of the kidney, on account of destructive disease of that organ, originally performed in 1861 by Dr. Charles L. Stoddard,[4] of Wisconsin, and by Peaslee in 1868, has since been successfully practised by Dr. J. T. Gilmore,[5]

---

[1] Am. Journ. Med. Sci., July, 1855, p. 272.
[2] Ibid., April, 1870, p. 365.
[3] New York Med. Rep., vol. iv., 2d Hex., 1807, p. 367.
[4] Med. and Surg. Rep. 1861, vol. vii. p. 126.
[5] Am. Journ. Obstetrics, May, 1871.

of Mobile, Alabama. The patient, who was pregnant at the time of the operation, was delivered at the full term, gestation having proceeded without any untoward occurrence. A *resumé* of all the recorded cases may be found in the *American Journal of the Medical Sciences* for January, 1873, page 277, and July, 1874, page 266.

The diagnosis and treatment of *perityphilitic abscess* have recently been ably illustrated by the researches of Dr. Willard Parker,[1] Dr. Gurdon Buck, Dr. Fordyce Barker, and Dr. Leonard Weber,[2] of New York. The operation of incising the muscular walls of the abdomen, or, in other words, evacuating the matter by a direct incision, was first performed, it would seem, in 1848, by Mr. Hancock, of London, and revived by Parker in 1867. By this procedure an immense amount of local and constitutional suffering is avoided, and the part and system are placed in a much better condition for ultimate recovery.

Professor Joseph Pancoast[3] is entitled to the credit of having performed the first successful plastic operation for the relief of exstrophy of the *bladder*. His patient was a young man from one of the Western States, who consulted him on account of this malformation in 1858. The operation consisted in forming a covering for the exposed surface by means of cutaneous flaps borrowed from the immediate neighbourhood, which were then inserted, and approximated at the middle line by suitable sutures. Excellent union occurred, and the man experienced great relief from his discomfort. but died at the end of two months and a half from an attack of pneumonia. A similar operation was performed soon afterwards by Dr. Ayres,[4] of Brooklyn, with equally gratifying results, his patient being a young woman.

---

[1] Med. Record, vol. ii. p. 25, and vol. xi. p. 12.
[2] New York Med. Journ., Aug. 1871, p. 142.
[3] N. A. Med.-Chir. Rev., July, 1859.
[4] Am. Med. Gaz., Feb. 1859.

Since that time the operation has been repeated, in a more or less modified form, by surgeons in different parts of the world. Dr. F. F. Maury,[1] of this city, has performed it three times, twice with admirable results; it has also been successfully performed by Dr. John Ashhurst,[2] by Dr. C. B. King, of Pittsburgh, and by Dr. Cheever and Professor Henry J. Bigelow,[3] of Boston; the latter of whom, in two instances, modified the procedure by removing the exposed mucous membrane of the bladder so as to bring the cutaneous flaps in direct contact with the raw surface, thus facilitating their adhesion. Both cases had a fortunate termination.

Under this head may be briefly described an interesting and ingenious operation, devised by Physick, and since repeatedly practised by Professor Pancoast and myself, for the cure of incurvation of the penis, a congenital malformation always occurring in association with hypospadias or epispadias. The operation consists in cutting away a V-shaped section of the cavernous bodies of the organ a short distance behind the corona, and approximating the raw surfaces with several points of suture. The amount of structure excised should be barely sufficient to straighten the penis. No skin is removed; and only a few vessels require ligation. The operation being completed, the organ is placed upon a gutta-percha splint, and covered with cold-water dressings, erections being prevented by the usual means. Nothing could be more ingenious than such a device, one founded upon the same principle precisely as the operation invented by Dr. John Rhea Barton for the cure of anchylosis. No mention, so far as my information extends, is to be found of this operation in any foreign works on surgery.

[1] Am. Journ. Med. Sci., July, 1871, p. 154.
[2] Ibid., p. 70.
[3] Boston Med. and Surg. Journ., Jan. 1876.

As *lithotomists*, American surgeons are not surpassed by any in the world. The success of the late Dr. Benjamin W. Dudley, of Lexington, is, if correctly reported, without a parallel. It must be observed, however, that, as he did not keep a record of his cases, altogether, I believe, 207, with a loss of only 6, that there must always remain a doubt upon this subject. The late Dr. James Bush,[1] who was, for many years, the assistant, and, if I mistake not, for a time a partner of Dr. Dudley, in giving an account of his operations, stated that these results were, as nearly as could be ascertained, correct. But even supposing that the mortality was considerably higher than the above figures tend to indicate, a fact of which I have not the slightest doubt, still the results would be sufficiently above the ordinary standard to show that Dudley was a great lithotomist. In looking over the second edition of my treatise on the *Urinary Organs*, issued in 1854, I find that there is a table of 895 cases by 21 different operators, including those of Dudley, with a mortality of 44, or 1 in $20\frac{1}{3}$; a degree of success which, to say the least, may justly be styled brilliant. It will not be without interest to add that in 426 of these cases the instrument used was the gorget, and the knife in 424, with a mortality for the former of 1 in $23\frac{7}{13}$, and for the latter of 1 in $19\frac{4}{11}$. Dudley invariably employed the gorget, an instrument now entirely disused on this side of the Atlantic. Valentine Mott always employed the knife, and, with the exception of Physick and Gibson, all the Philadelphia surgeons did the same. Dr. Alexander H. Stevens, Dr. N. R. Smith, Dr. Gurdon Buck, and several others have devised peculiar instruments for this operation, but whether they possess any superiority over the ordinary bistoury, now so much in use among surgeons, experience has not deter-

---

[1] Am. Journ. Med. Sci., Feb. 1838, p. 535, and April, 1846, 545. Trans. Am. Med. Assoc., vol. iv. p. 273.

mined. The method of operation usually selected is the lateral, as practised and perfected by William Cheselden, in the early part of the last century. A few of our surgeons prefer the bilateral method. The suprapubic operation was first performed in this country by Dr. William Gibson, of this city. The median operation, usually but improperly described as Allarton's method, has yielded good success in the hands of some of our surgeons, as Albert G. Walter, of Pittsburgh, and J. L. Little, and Thomas M. Markoe, of New York. The largest calculus ever successfully extracted from the bladder, although not without breaking it, weighed twenty ounces, the operator being Dr. A. Dunlap, of Springfield, Ohio, and the patient, who survived this ordeal nearly three years, a man sixty-six years old. The largest number ever removed was upwards of one thousand, varying from the size of a partridge shot to that of a bean, in the celebrated case of Chief Justice Marshall, who was cut by Physick when both were at an advanced age, the illustrious patient making an excellent recovery.[1]

It is generally supposed that lithotomy was first performed in this country about 1760, the operator being Dr. John Jones, subsequently professor of surgery at New York, and after his removal to Philadelphia for a time physician to Washington and Franklin. The latter, as is well known, laboured for many years under stone in the bladder, but was evidently afraid to submit to the use of the knife, notwithstanding the severity of his sufferings.

The most extensive collection of urinary calculi in the United States is in the museum of the College of Physicians of Philadelphia, and forms a part of the bequest of the late Professor Mütter. Many of the specimens, however, were obtained in Europe. To Professor Peter,[2] of Lexington,

[1] Am. Journ. Med. Sci., Feb. 1832, p. 537.
[2] Western Lancet, Nov. 1846, p. 241.

Kentucky, the profession is indebted for the most comprehensive series of analyses of urinary calculi yet made in this country. The specimens examined by him formed a portion of the collection contained in the museum of Transylvania University, as the result of the operations of Professor Dudley. The paper published upon the subject by Dr. Peter is well worthy of an attentive perusal. The largest stone ever successfully removed in this country was, as already stated, one which occurred in the practice of Dr. Dunlap, of Ohio. Professor Eve removed successfully one hundred and seventeen calculi from one patient, and Physick, as above mentioned, upwards of one thousand, all of course very small. The lateral operation has been most frequently performed by B. W. Dudley, Valentine Mott, N. R. Smith, J. P. Mettauer, Paul F. Eve, W. T. Briggs, and S. D. Gross; the bilateral by R. D. Mussey, P. C. Spencer, Willard Parker, Joseph Pancoast, and Paul F. Eve. Walter, Little, and Markoe have thus far been the principal advocates of the median method. The suprapubic and rectovesical operations have been performed only in a few instances, and are justifiable only under peculiar circumstances.

Lithotrity, more properly called lithotripsy, has never met with much favour on this continent, a circumstance so much the more surprising when it is remembered how common calculous diseases are in certain sections of the United States. The first operation of the kind was performed by Dr. Depeyre,[1] of New York, in 1830. This was followed soon afterwards by the cases of Alban Goldsmith, Jacob Randolph, William Gibson, Joseph Pancoast, N. R. Smith, J. Mason Warren, P. S. Spencer, Gurdon Buck, H. J. Bigelow, and others. To Randolph is usually, and I believe very justly, conceded the credit of having been the earliest and

---

[1] N. Y. Med. Journ., Feb. 1831, p. 383.

most successful cultivator of lithotrity in the United States. He operated a considerable number of times upon patients of all ages, and is said to have possessed an unwonted degree of dexterity in the use of the necessary instruments. Pancoast, N. R. Smith, Buck, and Bigelow, not to name others, also deserve honourable mention in connection with the subject.

For the relief of chronic cystitis, dependent upon hypertrophy of the prostate gland, one of the most painful and distressing of affections, attended with an almost constant desire to micturate, the late celebrated Mr. Guthrie, of London, suggested perineal lithotomy, an operation which, however, he never performed. This credit was reserved for Professor Willard Parker,[1] of New York, by whom the idea was first practically carried out in 1846. Five years later he repeated it in a similar case, the patient being an old man, who had long suffered from the effects of this disease, and who, consequently, was not a favourable subject for the operation. He was, nevertheless, greatly benefited by it, but died at the end of a month, apparently from uræmic poisoning. Since that period the operation has occasionally been repeated by other American surgeons, among others by Dr. Battey, of Georgia, and Professor Powell, of Chicago.

Of the great value of this operation, as a palliative measure, no longer any doubt can be entertained by any experienced or enlightened surgeon. That it will effect a permanent cure when the bladder has undergone serious structural lesion is, of course, impossible. In the milder cases, however, such a result may certainly be reasonably looked for in both sexes. The operation is less difficult than that for stone in the bladder, and almost the only

---

[1] New York Journ. of Med., July, 1851, p. 83; and Trans. N. Y. State Med. Soc., 1871, p. 345.

danger would probably be from hemorrhage, as the vessels about the prostate gland and neck of the bladder are always much enlarged and engorged in chronic catarrh.

A similar operation has occasionally been performed upon the female, as in the cases reported by Dr. N. Bozeman,[1] and by Professor T. G. Richardson. Vaginal cystotomy for the cure of this affection was originally suggested by Dr. J. Marion Sims, in 1858, but was, if I mistake not, first performed by Dr. T. A. Emmet.[2] The wound in the bladder, in this operation, is kept open until complete relief is afforded, when, if necessary, it is artificially closed. Dr. T. W. Howe[3] recently successfully treated an obstinate case of chronic cystitis by dilatation of the neck of the bladder. The value of vaginal cystotomy, like that of perineal cystotomy in the male, is now fully established, and much of the interest which the procedure has elicited is due to American genius and enterprise.

The treatment of *strictures* of the urethra is a subject which has received much attention on this side of the Atlantic. Many ingenious instruments have been devised as guides to the bladder, and for overcoming the obstruction; instruments, perhaps, not always so valuable as the claims set up for them by their respective authors would seem to indicate. The names of Van Buren, Gouley, and Otis deserve special mention in connection with this subject. In regard to the perineal section, or external urethrotomy, for the cure of tight, callous strictures attended with fistulous openings in the perineum and scrotum, the operation, originally described by Desault, was well known in the United States, and frequently performed, before a knowledge of it was introduced to the notice of European surgeons by the

---

[1] N. Y. State Med. Soc. Trans., 1871, p. 326.
[2] Am. Practitioner, Feb. 1872, p. 65.
[3] Medical Record, vol. x. p. 550.

late Professor Syme, of Edinburgh, the principal operators having been Physick, Stevens, Jameson, Rogers, and Warren.

The fact that *masturbation*, when carried to great excess, may, in certain constitutions and states of the system, give rise to insanity has long been known to medical men, and various remedies have accordingly been suggested for its cure. Of the purely medical treatment this is not the place to speak, and of the surgical there are only two methods which need here be mentioned. I allude to castration and the excision of a portion of the deferent tubes. Of these two proceedings, both practised by American surgeons, the former has met not only with severe criticism, but pointed condemnation. That such an operation is, at first sight, well calculated to elicit such rebuke is not surprising; I was myself at one time an uncompromising opponent of it, and it has only been after much reflection and deliberation that I have reached a different conclusion. No man who is in his right senses would think of employing such a remedy in ordinary cases; but when the habit of onanism is so firmly established as completely to wreck the patient's mind, and all other remedies have been employed in vain, surely any means calculated to rescue the unfortunate sufferer from destruction is not only justifiable but in the highest degree proper. The remedy, it will be perceived, is a very different one from excision of the clitoris, so absurdly practised by Baker Brown and others, for the cure of masturbation in the female. By removing the testicles the surgeon strikes at once at the very root of the evil, and thus places the patient in a much better condition for the favourable action of other remedies and his ultimate recovery. Professor Van Buren,[1] of New York, in 1848, in a case of partial idiocy

---

[1] Genito-Urinary Diseases, Van Buren & Keyes, New York, 1874, p. 458.

consequent upon excessive onanism, had recourse to the excision of a section of each vas deferens, as a substitute for castration, all other means having failed to break up the disgusting habit. The operation, however, proved of no advantage. I am not aware that this operation, which, of course, is eventually followed by complete atrophy of the testes, has been repeated either in this country or in Europe.

The *extirpation of tumours*, even of the most simple kind, is frequently attended with much difficulty, especially in the hands of young and inexperienced surgeons; a difficulty which is often greatly enhanced when the morbid growth is embedded in important structures, as in the cervical, axillary, inguinal, femoral, and popliteal regions. At the present day rules for meeting these contingencies abound in our surgical treatises; but it was far different a third of a century ago, when the only great work on surgery in the English language was Samuel Cooper's Dictionary; a work of world-wide fame, reprinted in this country, and translated into most of the languages of Europe. It was about this time that Dr. Alexander H. Stevens,[1] of New York, published his celebrated clinical lecture on the operative surgery of tumours, delivered at the New York Hospital, in which he enunciated these two golden rules as guides in all operations of this kind: First, to cut down fairly upon the morbid growth before commencing its dissection; and, secondly, to remove the diseased mass and nothing more. Of the value of these suggestions every modern surgeon is fully aware, and it is hardly necessary to add that, as they were the result of a large clinical experience, they were received with great respect and consideration by the medical profession both at home and abroad.

The huge growths, which are occasionally developed during the progress of elephantiasis of the skin and connec-

---

[1] Boston Med. and Surg. Journ., vol. xxii. p. 53.

tive tissue in the lower extremities, in the genital organs, and in other parts of the body, although uncommon in this country, are now and then met with, either among our own people, or in persons sent hither for advice and treatment from South America and the West Indies; and several cases have occurred in which large masses of this kind seated in the scrotum have been removed by some of our surgeons, as Picton, of New Orleans, N. Bozeman and J. S. Thebaud, of New York, and John Neill, of Philadelphia. The weight in two of these cases, both successful, was, respectively, fifty-three and fifty-one and a half pounds. Dr. Bozeman's patient died of peritonitis nearly a fortnight after the operation. The tumour weighed forty pounds.

With a view of arresting the growth in elephantiasis, Dr. Carnochan,[1] of New York, in 1851, was induced to tie the main artery leading to it, hoping, by cutting off its supply of blood, to cause the growth to shrink, or to undergo more or less atrophy. His case was one of elephantiasis of the lower extremity, for the relief of which he ligated the femoral artery. In another instance—one of elephantiasis of the head, face, ears, and neck—the same surgeon tied both the primitive carotid arteries after an interval of six months.[2] Since Dr. Carnochan's first case, ligation of the femoral artery for the relief of elephantiasis of the lower extremity has been practised by Dr. Campbell, formerly of the Philadelphia Hospital, Dr. Thomas G. Morton, of this city, Dr. Bauer, of St. Louis, Dr. McGraw, of Detroit, and several others. In no instance, however, I believe has a complete or permanent cure followed these laudable efforts; a significant fact not to be overlooked in estimating the true value of this procedure, the more especially as most of the operators are men of acknowledged ability. In a case of ele-

[1] N. Y. Journ. Med., Sept. 1852, p. 161.
[2] Am. Journ. Med. Sci., July, 1867, p. 109.

phantiasis at the Pennsylvania Hospital, under the care of Dr. Addinell Hewson, marked relief was afforded by that gentleman's so-called earth dressing, aided by the use of the roller.

In *excisions* of the joints and jaws, and in operations upon the bones for the removal of sequesters, as well as in the extirpation of tumours, surgeons in all parts of the world have long been in the habit of employing what is called the *curvilinear incision*, the first application of which was at one time a subject of considerable controversy, especially between Velpeau and Mott. In his great work on "Operative Surgery," the former of these distinguished men dwells with special emphasis upon this form of incision, and in a letter addressed to Dr. Mott, in the American translation of the work, by Dr. Townsend, he observes: "A point which I deem important is that which relates to my *new* processes for the extirpation of tumours, the amputation of the jaws, and in exsections. Persuaded that you must perceive at a glance all the advantages to be derived from the curved incision, substituted for the straight, in the extirpation of tumours which may be removed without trenching upon the integuments, I will make no further remarks upon the subject in addition to those you will find in my notice."

In his comments upon these remarks, Professor Mott states that he had invariably, in all his operations for the removal of the lower jaw, since 1821, employed the curvilinear incision, a procedure to which, he continues, his French confrere justly attaches such great importance. "His description," he further adds, "of its advantages and superiority over every other mode of reaching the osseous structures to be exsected by the saw or nippers, is so clear and graphic that I have nothing to add to it whereby I could impress upon the mind of the surgeon its decided preference over every other mode." It thus clearly appears that our countryman is fairly entitled to the credit of priority in this important matter. I may state that I employed the curvilinear incision in many of my operations long before I became acquainted with the fact that it had been used by Dr. Mott, and that Dr. Joseph Pancoast and other Philadelphia surgeons have, on all suitable occasions, availed themselves of the curvilinear incision for upwards of a third of a century. Of its great advantages every operator is fully aware.

Excision of nerves for the cure of neuralgia and other affections has been frequently practised by American surgeons, and has, especially of late, been carried to a very great extent, in one case embracing the removal of a section of the entire brachial plexus. The superior maxillary nerve has been cut out beyond the ganglion of Meckel, Dr. Carnochan,[1] of New York, having led the way in this bold procedure, in which he has been followed by Pancoast, Blackman, W. H. Mussey, and several others. A very ingenious operation for exposing and exsecting the second and third branches of the fifth pair of nerves was devised, many years ago, by Professor Pancoast, and is fully described in the second volume of my *System of Surgery*, issued in 1872. In the four cases in which it was performed, the cure, in all, was permanent, after years of agonizing suffering. An instance was recently reported by Dr. J. L. Stewart, of Erie, Pennsylvania, in which he exsected three inches of the median nerve for the relief of neuralgia caused by a gunshot wound of the forearm inflicted six years previously. In 1873, Dr. F. F. Maury[2] exsected the brachial plexus of nerves from an elderly man, an inmate of the Philadelphia Hospital, on account of the exquisite suffering occasioned by numerous neuromatous tumours occupy-

---

[1] Am. Journ. Med. Sci., Jan. 1858, p. 134.
[2] Ibid., July, 1874, p. 29.

ing the integument of the shoulder, back, and chest. After the failure of all other known and approved methods of treatment, the brachial plexus was exposed, and a section, embracing four-fifths of an inch of nerve substance, removed. The patient, who gradually recovered from the effects of the operation, experienced great relief immediately after it was over, the improvement lasting for a few months, but the neuralgia has of late been again severe. An operation of a similar kind, but not for the same disease, has been performed by Drs. Sands and Seguin, and is reported in the New York *Archives of Scientific and Practical Medicine* for 1873. I have myself done a considerable share of this kind of work; and some interesting examples of a rather formidable character were reported by the late Dr. Josiah C. Nott,[1] of Mobile. In a remarkable case of vaginal neuralgia of twelve years' duration, in the care of Dr. Thomas G. Morton,[2] of Philadelphia, a speedy and permanent cure was effected by the excision of the perineal nerve. The nerve was exquisitely tender on pressure, and rolled under the finger like a firm cord.

Although it is not in accordance with the design of this paper to speak of works on surgery, I cannot forbear referring here to two monographs, the productions of three American authors, inasmuch as they have a direct bearing upon the subject under discussion. I allude to the work of Dr. S. Weir Mitchell, Dr. George W. Morehouse, and Dr. William W. Keen, on *Gunshot Wounds and other Injuries of Nerves;* and to the admirable monograph on *Injuries of Nerves and their Consequences,* by Dr. S. Weir Mitchell, issued, respectively, in 1864 and 1872. The publication of these productions may be considered as forming a new era in the history of these lesions. Up to that period the whole

---

[1] Bone and Nerve Surgery, Nott, Phila., 1866.
[2] Am. Journ. Med. Sci., Oct. 1873, p. 399.

subject was involved in mystery, and it was only by studying it in the light of clinical experience that it was successfully unravelled. The work of Dr. Mitchell is founded upon the careful observation and analysis of several hundred cases of injuries of nerves, and upon numerous experiments performed on animals, with a view of determining the physiological questions of the influence of pressure on nerves, elongation, and separation. Of this treatise it is not too much to say that it constitutes one of the most valuable contributions made to the medical and surgical science of the present day. Ever since the date of its publication the lesions of nerves have been regarded from a new standpoint.

As intimately connected with the condition of the nervous system, brief reference may here be made to another field of investigation, in which the labours of Dr. Mitchell have been conspicuously displayed. I allude to what is called *snakebite*, a subject concerning which science had been silent since the days of the Abbé Fontana in the latter part of the last century. In 1860, Dr. Mitchell gave to the profession the results of three years of work, almost unremittingly devoted to the object, pointing out in the clearest manner the nature, accidents, and incidents of venomous wounds, followed, the ensuing year, by a full account of their true treatment, up to that period little, if at all, understood. Although other American physicians and scientists had performed experiments with the poison of the rattlesnake, it remained for the Philadelphia philosopher to throw an amount of light upon the subject, which has hardly a parallel in any other branch of scientific investigation. I will only add that the researches of Dr. Mitchell have been embodied in a quarto volume, issued by the Smithsonian Institution, at Washington City, in 1861.

Plastic surgery has received much attention in this country, especially during the last thirty years. With the cultivation of this branch of the healing art are associated the names of a number of our most distinguished operators. Mention may be made, more particularly, of those of Joseph Pancoast, T. D. Mütter, John Watson, J. Mason Warren, F. H. Hamilton, Gurdon Buck, and David Prince, who have all rendered good service in this way. Pancoast[1] is entitled to great credit for his rhinoplastic operations, for the successful performance of which he has devised a special suture admirably adapted to secure effective union of the flaps, and generally known by his name or as the plastic suture. As this suture is described in all our native treatises on surgery, it need not detain us here. European surgeons seem to be ignorant of its merits, if, indeed, not of its existence. The Indian method is the one usually adopted in this country for the restoration of lost or mutilated noses, and for closing gaps in the cheeks and lips. J. Mason Warren[2] occasionally adopted the process of Tagliacozzi, but, instead of raising the skin of the arm, as in the original method, he borrowed the requisite amount of substance from the forearm, a short distance above the wrist, thus rendering the confinement of the limb much less irksome. Dr. Gurdon Buck has recently obtained some remarkable results by the sliding process in operations for the cure of deformities of the nose, lips, and cheeks.

Dr. John Watson,[3] of New York, in 1844, by a plastic operation closed a large opening in a man's forehead, caused by a syphilitic ulcer, this being probably the first case of the kind thus treated in this country. The first instance in which the cure of an ulcer was effected by grafting healthy integument upon its raw surface occurred in the practice of

---

[1] Operative Surgery, Phila., 1844, p. 345, and Med. Ex., N. S., vol. vii. p. 238.

[2] Surgical Obs., with Cases, Boston, 1867.

[3] Am. Journ. Med. Sci., Oct. 1844, p. 537.

Professor F. H. Hamilton, then of Buffalo, the patient being a man who had lost a large portion of the skin of the leg by the fall of a heavy stone; the integument was borrowed from the sound limb, but was wholly inadequate to cover the ulcer. The graft gradually expanded centrifugally, and at the end of ninety days the cicatrization was complete. "Since the date of this observation," remarks Professor Hamilton, "I have repeated it many times with almost uniform success." A full account of this case will be found in the *New York Journal of Medicine* for 1854. Whether Mons. Reverdin borrowed his ideas of skin-grafting from the proposal of the New York surgeon, enunciated in 1847, in the *Buffalo Medical and Surgical Journal*, before he had an opportunity of putting it into practice, is a question which need not detain us here. The principle of the two processes is precisely alike.

Mütter and Pancoast, as, indeed, other American surgeons, have reported a number of instances in which plastic operations on a most extensive scale were performed for the relief of the unsightly and injurious contraction caused by the cicatrices left by burns and scalds. Much credit is justly due to these two gentlemen for the boldness of their undertakings and for the very able manner in which they were executed. Prior to the publication of the results of their cases, little attention had been bestowed upon this particular branch of the subject by our European brethren. The late Mr. Prigden Teale, of Leeds, it is believed, took the lead in this class of cases in England.

Ophthalmology, as a specialty, or separate branch of surgery, took its rise in this country upwards of half a century ago in connection with the establishment of Eye and Ear Infirmaries in nearly all the larger cities of the Union, conducted by men who considered themselves as peculiarly fitted for this kind of work.

The New York Eye and Ear Infirmary was opened in

1822, the Pennsylvania Eye and Ear Infirmary at Philadelphia in 1822, the Massachusetts Eye and Ear Infirmary at Boston in 1829, and Wills Ophthalmic Hospital of Philadelphia in 1834. The American Ophthalmological Society was instituted in 1864, and in its transactions may be found numerous proofs of original and valuable contributions by its members. Dr. E. Williams, of Cincinnati, was, if I mistake not, the first physician on this continent who made ophthalmology an exclusive specialty, and who delivered the first regular course of lectures on ophthalmic medicine and surgery in a medical school.

The operations performed upon the eye during the first half of the present century were limited almost exclusivsly to the cure of cataract, the formation of artificial pupil, and the evacuation of fluids and of foreign matter from the chambers of the eye. Extraction of cataract was an infrequent procedure, confined to the hands of a few skilful men; comminution and depression were the common operations. Occasionally an eye was extirpated on account of malignant disease, but of enucleation of the ball as now practised we were in total ignorance. The operation for the cure of strabismus was unknown. In all these and other respects American surgeons did neither less nor more than their European brethren. The invention of the ophthalmoscope by Helmholtz, in 1852, revolutionized the science and the art of ophthalmology; and, if we have had no share in the vast improvements incident to that invention, one of the grandest of the age, we have kept steady pace with the advances made abroad, and, perhaps, added a few things to the general stock of knowledge, not wholly devoid of value.

Among the more important of these contributions may be mentioned the needle-knife of Dr. Isaac Hays,[1] long successfully employed by him, as well as others, for the com-

---

[1] Am. Journ. Med. Sci., July, 1855, p. 81.

minution of cataract; the use of a suture in the cornea after the flap extraction of cataract by Dr. Henry W. Williams,[1] of Boston; the production of cataract in frogs by the administration of saccharine substances, by Dr. S. Weir Mitchell,[2] of Philadelphia; the laceration of opaque capsule, by Dr. Agnew, of New York, with an instrument combining the principle of a needle and hook, his method of removing foreign bodies from the cornea, and, lastly, his improvements in canthoplasty and the operation for divergent strabismus; the gymnastic treatment of asthenopia proposed by Dr. Dyer, of Pittsburgh; the test cards of Dr. Green,[3] of St. Louis, and of Dr. Pray,[4] of New York, for the diagnosis of astigmatism; the wire loop of Dr. Levis for facilitating the extraction of cataract; the improved ophthalmoscope of Dr. Knapp[5] and of Dr. Loring,[6] of New York; the operation of Dr. Knapp for the removal of tumours from the optic nerve without sacrificing the eye, his hook for extracting foreign matter from the ball,[7] and, lastly, his improvement of the ring forceps;[8] the theory of Dr. Loring[9] of the cause of the light streak in the vessels of the retina, and of the halo around the fovea; the optometer of Dr. Thomson,[10] of Philadelphia, for the diagnosis of ametropia, together with his improvements in the use of cylindrical glasses in conical cornea, and his discovery of the connection between astigmatism and posterior staphyloma; finally, we must not forget to allude, in terms of high commenda-

---

[1] Trans. Am. Ophthal. Soc., 1866 and 1867, p. 58.
[2] Am. Journ. Med. Sci., Jan. 1860, p. 106.
[3] Ibid., Jan. 1867, p. 117.
[4] Archiv Opthal. and Otol., vol. i. p. 17.
[5] Trans. Am. Ophthal. Soc., 1873, p. 109.
[6] Am. Journ. Med. Sci., April, 1870, p. 340.
[7] Trans. Am. Ophthal. Soc., 1873, p. 105.   [8] Ibid.
[9] Ibid., 1870, p. 122, and 1873, p. 87.
[10] Am. Journ. Med. Sci., Jan. 1870, p. 76, and Oct. 1870, p. 414.

tion, as the very latest American addition to ophthalmology, to the improved ophthalmoscope of Dr. Shakespeare, of Philadelphia, published in the *American Journal of the Medical Sciences* for Jan. 1876, a contrivance, apparently, admirably adapted to physiological experiments as well as to the diagnosis of diseases of the eye.

In the American appendix to Cooper's Dictionary of Surgery, issued in 1842, appears the curious statement by the editor, Dr. David Meredith Reese, that the straight muscles of the eye were divided by Professor William Gibson, of Philadelphia, for the cure of strabismus several years before the operation was performed by Dieffenbach, of Berlin. The result, however, it would seem, was not satisfactory, and the operation was finally, at the earnest solicitation of Dr. Physick, abandoned. After a patient research, I am unable to find any allusion to this operation in the writings of the Philadelphia surgeon; and the assertion of Dr. Reese appears to be based wholly upon a statement of the late Dr. Alexander H. Hosack, of New York, who, it is alleged, heard Dr. Gibson refer to the subject in his lectures in the University of Pennsylvania. In this country the operation of Dieffenbach was first performed by Dr. Willard Parker, and soon afterwards by Dr. Isaac Hays, of Philadelphia, Dr. Detmold, of New York, Dr. Golding, of Arkansas, and by myself, then a resident of Kentucky.

Although *Aural surgery* had no existence in this country as a scientific study until a quarter of a century ago, it may confidently be affirmed that it is nowhere cultivated with more ardor, or practised with more success, than it is at present on this side of the Atlantic. If we have not made any advances or discoveries, our knowledge of the pathology, diagnosis, and treatment of the diseases and injuries of the ear is of the highest order, fully on a level, in every respect, with our knowledge of the other branches of the healing art.

Among the earliest and most valuable contributions to

otology may be mentioned the researches of Dr. Edward H. Clarke, late Professor of Materia Medica at Harvard University, "On the Causes, Effects, and Treatment of Perforation of the Membrana Tympani," published in the *American Journal of the Medical Sciences* for January, 1858. This paper was followed in the same year by that of Dr. Laurence Turnbull, in the *Philadelphia Medical and Surgical Reporter*, "On the Treatment of Otorrhœa considered as a sequel of Scarlatina." The same writer, in 1862, published a valuable memoir on "Otitis Interna," illustrated by three cases treated by perforation of the mastoid cells, the first operations, it is believed, of the kind performed in this country.

Among the more recent contributions to the subject is that of Dr. Edward H. Clarke, of Boston, on the Nature and Treatment of Polypus of the Ear; of Dr. Charles H. Burnett, of Philadelphia, on the Mechanism of the Ossicles of hearing and the Membrane of the Round Window; of Dr. Hermann Knapp, of New York, on the Inflammations of the Internal Ear; of Dr. J. Green, of St. Louis, on the Action of Condensed Air on the Eustachian Tube; of Dr. J. O. Green, of Boston, on Otorrhœa; of Dr. Seeley, of Cincinnati, on Aural Polypus and its Treatment; of Dr. Rumbold, of St Louis, on the Functions of the Eustachian Tube in relation of the renewal and density of the air in the Tympanic Cavity; and of Dr. A. H. Buck, of New York, on the Mechanism of Hearing. For further information as to what has been done in recent years in this department of surgery, I must refer the reader to the Transactions of the American Otological Society.

I cannot omit here the insertion of a brief account of the *ward carriage*, as it is termed, of Dr. Thomas G. Morton, of Philadelphia, devised a few years ago, and employed with such signal benefit at the Pennsylvania Hospital, for which it was originally constructed.[1] Why it is not in general use in all such institutions is a question that might well be asked. Its arrangements are so perfect, and its conveniences so great, that it is only surprising it was not thought of long ago. The appliances for cleansing wounds and ulcers are so complete that there can no longer be any excuse for an impure or infectious condition of the atmosphere in the wards of any hospital in which the carriage is employed.

The same ingenious surgeon, in 1872, invented an admirable *hospital carriage* and bed-elevator for the transfer of surgical and medical cases, from one ward to another, or to and from the amphitheatre for operation and clinical instruction. By this contrivance the inconveniences attendant upon the ordinary procedure of carrying the patient about upon his bed, a mode of conveyance often accompanied with great jarring and discomfort, are entirely obviated. The apparatus consists of a double truck, the upper one of which is elevated by a series of cams, running upon a narrow iron track, and capable of being raised and depressed at pleasure by means of a chain worked by a crank. It is six feet and five inches in length by two feet and nearly a half in width, is constructed of oak, and is moved by four wheels supported by steel pins. A description of the apparatus, accompanied by a wood-cut, may be found in the *American Journal of the Medical Sciences* for July, 1874.

To write the history of a century of American Surgery without a recognition of the services rendered by our *Army Surgeons*, would be an unjustifiable omission. My only regret is that my limited space does not permit me to enter more into particulars.

Of the amount of service contributed during the late war by the medical profession a tolerably correct estimate may

---

[1] Am. Journ. Med. Sci., Jan. 1867, p. 135.

be formed when it is stated that 11,608 surgeons and assistant surgeons were on active duty either in the field, in the camp, or in the hospital. Of this number, 9 were killed by accident, and 32 in battle, by guerillas, or by partisans; 83, of whom 10 died, were wounded in action; 4 died in prison, 7 of yellow fever, 3 of cholera, and 271 of various diseases, mostly incident to military life; presenting thus an aggregate of 409 names. The amount of labour performed by the medical staff during the war may be faintly guessed when it is stated that 157,423 cases of wounds and diseases occurring among the white troops were treated in general hospitals alone, exclusive of the vast numbers that were attended in regimental and post hospitals.

Surgeon-General Barnes, at the close of his annual report for 1865, in speaking of the character of the medical staff of the army, says: "I desire to bear testimony to the ability, courage, and zeal manifested throughout the war by the officers of the Medical Department under all circumstances, and upon all occasions. With hardly an exception they have been actuated by the highest motives of national and professional pride, and the numbers who have been killed and wounded bear most honourable testimony to their devotion to duty on the field of battle."

How far the services of these faithful and devoted men were instrumental in bringing the war to a favourable issue, and in rescuing the lives of wounded and diseased soldiers by the timely and judicious interposition of their scientific knowledge and skill, is, of course, merely a matter of conjecture. That they were of vast importance no intelligent man will doubt; and it is not too much to say that the country owes the Medical Staff a great and lasting debt of gratitude.

Most of the hospitals erected during the war were constructed on the pavilion principle, and were supplied with every possible accommodation, comfort, and convenience.

It seemed, indeed, as if the Government had been determined, through its medical staff, to rob injury, disease, and surgical operations of their terror, such was the palatial character of many of these establishments. Of these hospitals upwards of two hundred were still in existence on the 1st of January, 1865. "The hospital transport system included four first-class sea-going steamers, equipped with stores and supplies for five thousand beds, besides a large number of river hospital boats, hospital railway trains, ambulance, etc."

The *Army Medical Museum* at Washington, initiated by Surgeon-General Hammond, and perfected by Surgeon-General Barnes and his medical staff, under the auspices of the Government, is confessedly the noblest institution of the kind in the world. In 1873 it contained upwards of fifteen thousand catalogued specimens, comprising objects in surgery, medicine, anatomy, microscopy, and comparative anatomy, plaster casts, drawings, crania of Indians, skeletons and crania of animals, birds, reptiles and fishes, and a complete collection of models of ambulances, litters, and other appliances for the transportation of the sick and wounded, artificial limbs, and photographs illustrative of surgical operations; in short, everything that can impart completeness to such an establishment.

In the same edifice is the *Library of the Surgeon-General's Office*, embracing 38,000 volumes, and 40,000 pamphlets, making the total number of titles nearly 70,000. About 2000 of the works are of a non-professional character, illustrative of the history of the late war, meteorology, physics. and other subjects. The library is especially rich in its collection of American medical periodicals. Much of the success which has attended the establishment of this library— certainly one of the largest and finest of the kind in the world—is justly due to the industry, zeal, and judgment of Assistant-Surgeon John S. Billings, who, amidst other arduous duties, has devoted much of his time to the enterprise, and has, in addition, found leisure to prepare an admirable catalogue of the authors of the books, with an alphabetical index of subjects, thus greatly facilitating the labour of reference on the part of those who desire to consult the collection.

To Surgeon-General Barnes and to two of his assistants, Dr. George A. Otis and Dr. Joseph J. Woodward, U. S. Army, the country is indebted for their unceasing efforts to build up a great Army Medical Museum, and for the vast labour, talent, learning, judgment, and enterprise which they have displayed in publishing the surgical and medical memoirs of the late war; works which, when completed, will reflect imperishable lustre upon the medical staff of the army, and upon our national medical literature. Dr. Billings, too, deserves honourable mention, in connection with the Surgeon-General's Bureau, for the ability which marks his reports upon hygiene of the army, barracks, hospitals, and other subjects.

Statistics, illustrative of the results of surgical operations, diseases and accidents, are, as is well known, often very troublesome and laborious undertakings, especially when attempted upon a large scale. Of this kind of work, unknown a third of a century ago, a very considerable amount has been done in this country, and, for the most part, done well. Among the more important and laborious of these contributions may be mentioned those of George W. Norris,[1] Stephen Smith,[2] Willard Parker,[3] and Thomas G. Morton[4] on the ligation of arteries for the cure of aneurism, the suppression of hemorrhage, and the arrest of morbid growths;

---

[1] Am. Journ. Med. Sci., N. S., vols. x., xiii., xiv.
[2] New York Med. Journ., Sept. 1852, p. 184, and Jan. 1853, p. 9.
[3] Ibid., Nov. 1852, p. 307.
[4] Penna. Hosp. Rep., vol. i. 1868.

of S. W. Gross,[1] on the treatment of aneurism by digital compression; of George Hayward,[2] George W. Norris,[3] and James R. Chadwick[4] on amputations in general; of Stephen Smith,[5] George C. Blackman,[6] and Thomas G. Morton[7] on amputations of the hip-joint; of John H. Brinton on amputations of the knee-joint;[8] of George W. Norris, on ununited fractures[9] and compound fractures;[10] of Blatchford, Spoor, and J. L. Smith on hydrophobia; of Lewis A. Sayre,[11] Chas. K. Winne,[12] and John Ashhurst[13] on excision of the hip-joint; of Stephen Rogers[14] on excision of the scapula; of Richard M. Hodges,[15] in an admiraable monograph, on excisions in general; of Frank H. Hamilton[16] on deformities after fractures; of John S. Billings[17] on trephining in epilepsy; of S. D. Gross on foreign bodies in the air-passages;[18] of Gurdon Buck[19] on laryngeal tumours; of J. Mason Warren and Frank H. Hamilton on the results of surgical operations in malignant diseases; of S. W. Gross on the treatment of compression of the brain

[1] North Amer. Med.-Chir. Rev. 1859.
[2] Am. Journ. Med. Sci., N. S., vol. i.
[3] Ibid., vol. xxii., and N. S , vol. i.
[4] Boston Med. and Surg. Journ., 1872, vol. ix., appendix.
[5] N. Y. Journ. Med., N. S., vol. ix.
[6] Western Lancet, vol. xviii. p. 7.
[7] Am. Journ. Med. Sci., July, 1866.
[8] Ibid., April, 1868, p. 305.
[9] Ibid., Jan. 1842, p. 13.
[10] Contributions to Practical Surgery, Norris, Phila., 1873.
[11] Trans. Am. Med. Assoc., vol. xiii. p. 469.
[12] Am. Journ. Med. Sci., July, 1861, p. 26.
[13] Penna. Hosp. Rep., vol. ii. p. 143.
[14] Am. Journ. Med. Sci., Oct. 1868, p. 359.
[15] Boston, 1861.
[16] Trans. Am. Med. Assoc. 1855-56-57.
[17] Am. Journ. Med. Sci., July, 1861, p. 299.
[18] Phila., 1854.
[19] Trans. Am. Med. Assoc., vol. vi, 1853, p. 507.

as met with in army practice;[1] of George A. Otis and S. W. Gross on amputations and excisions necessitated by gunshot wounds;[2] of John Ashhurst on injuries of the spine;[3] of S. W. Gross on ligation of the veins;[4] of Thomas G. Morton on amputations,[5] and on lithotomy and lithotrity at the Pennsylvania Hospital;[6] of S. W. Gross on synostosis of the knee-joint;[7] of William Hunt[8] on fractures of the larynx; of Thomas H. Andrews[9] on penetrating wounds of the skull; of R. O. Cowling on tetanus; of J. Solis Cohen[10] on croup in its relations to tracheotomy; of G. A. Van Wagenen on fractures treated at Bellevue Hospital with plaster of Paris apparatus; and of Charles W. Dulles[11] on suprapubic lithotomy.

Finally, let it not be supposed from what precedes that the American surgeon is a mere operator; if he ranks high in this particular, he ranks high also as a therapeutist. Nowhere, it may safely be asserted, are the great principles of surgery better taught, or better understood, than they are in this country. As a general practitioner, skilled in diagnosis, and in the art of prescribing, it is no presumption to affirm that he has no superior. If, as a body, we are deficient in any particular, it is in the more refined and subtile portions of our studies; studies which, after all, are of no essential practical importance, and which, it is not too much to say, will in due time receive their just proportions.

[1] Am. Journ. Med. Sci., July, 1873.
[2] Ibid., Oct. 1867, p. 423.
[3] Injuries of the Spine, Phil. 1867.
[4] Am. Journ. Med. Sci., Jan. and April, 1867.
[5] Ibid., Oct. 1870.
[6] Penna. Hosp. Rep., vol. ii. 1869.
[7] Am. Journ. Med. Sci., April, 1868.
[8] Ibid., April, 1866.
[9] Penna. Hosp. Rep., vol. i. p. 281.
[10] Trans. Med. Soc. of Pennsylvania, 1874.
[11] Am. Journ. Med. Sci., July, 1875, p. 39.

As a proof of what has been here stated, it is only necessary to refer to a few well-known facts. Thus, the treatment of wounds and injuries has been greatly simplified during the last fifty years. The importance of rest and of the prevention of pain in these and other lesions is universally recognized. The adhesive process is aimed at after all operations, whether small or great. None but the most simple dressings are employed. Little, if any faith, is placed by any enlightened or experienced surgeon on this side of the Atlantic in the so-called carbolic acid treatment of Professor Lister, apart from the care which is taken in applying the dressing, or, what is the same thing, in clearing away clots and excluding air from the wound;—an object as readily attained by the "earth dressing" of Dr. Addinell Hewson,[1] of Philadelphia, and by the oil dressing—composed of a thin layer of cotton or patent lint, wet with olive-oil—which I have myself employed for many years, with signal benefit, in nearly all cases of wounds under my charge, whether the result of accident or design. Such a covering, at once light and simple, answers every purpose, even in the largest wounds, excluding the ingress of foreign matter, and keeping the tissues moist and comfortable. The treatment of fractures with the aid of adhesive plaster and other appliances has received, as we have seen, some of its most valuable improvements in this country. No more beautiful or delicate instruments are manufactured than in Philadelphia, New York, Boston, and other large cities of the Union. The American artificial limb is celebrated everywhere for its elegance, lightness, and durability. Our adhesive plaster was formerly, if, indeed, it is not still, unequalled in excellence. Collodion, as a surgical dressing, owes its origin to an American physican, Dr. J. P. Maynard, of Boston. The wire suture was introduced into

[1] Earth as a Topical Application in Surgery, Phila., 1872.

regular practice by Dr. J. Marion Sims; and Professor Joseph Pancoast, many years ago, invented an ingenious suture, of peculiar construction, admirably adapted to promote the success of rhinoplastic and other plastic operations. The use of compressed sponge, so valuable in the treatment of many surgical affections, is mainly of American origin, largely due to the labors of the late Dr. J. P. Bachelder, of New York. The pathology and treatment of inflammation are much better understood; a much more rational system of dietetics prevails; inordinate depletion, in the form of heroic bleeding and purging, has ceased to be the order of the day; the influence of Nature in the cure of disease is better appreciated; and sanitary science has pointed out new paths of inquiry and of investigation, all tending to prevent suffering and to prolong life. Two most valuable remedies, of native origin—gelseminum and veratrum viride—have been added to the surgeon's pharmacopœia, and hold the same exalted rank in this country as depressants in the treatment of inflammation, neuralgia, and rheumatism, that aconite enjoys in Great Britain and on the Continent of Europe.

Conservative surgery is nowhere more thoroughly appreciated than it is in this country. Taking its rise with Physick, who was a great advocate of rest in the treatment of local diseases, and who scrupulously refrained from the employment of the knife whenever it was possible, it has always formed a prominent trait in the conduct of every enlightened American practitioner. Comparatively few knivesmen, properly so called, exist among us, and it is worthy of note that their career is usually as shortlived as it is inglorious.

It will thus be seen that during the hundred years which have just elapsed the medical profession has kept steady pace with the general progress of the arts and sciences on this continent; and not the least gratifying circumstance connected with it is the knowledge that it occupies a position in the social circle not accorded to it in any other part of the world, Great Britain, perhaps, excepted. The cultured and refined American physician is a prince among men. Let us be grateful for what we are and for what we have done; grateful that the past has such a splendid record—that it has left such distinct footprints upon the sands of Time—and that the future is so full of bright promises. The surgeons who are now rapidly passing away—those links between the past and the present—who have so long borne the heat and burden of the day—are not afraid to entrust to their younger brethren the sacred duty of carrying on the work which they have so zealously laboured to advance and improve. They feel assured that the honour, the dignity, and the glory of American surgery will be safe in their keeping, and that the century closing with the year 1976 will open for medicine one of the brightest pages in the history of human progress.

OBSTETRICS AND GYNÆCOLOGY.

BY

T. GAILLARD THOMAS, M.D.,

PROFESSOR OF OBSTETRICS AND DISEASES OF WOMEN AND CHILDREN, IN THE
COLLEGE OF PHYSICIANS AND SURGEONS, NEW YORK.

OBSTETRICS AND GYNÆCOLOGY.[1]

THE progress of philosophy, theology, politics, and science has never, in the history of the world, been marked by steady, monotonous, and gradual advancement. For long periods it has appeared to be so, but now and then, once in a century perhaps, each of these departments has felt the impetus imparted to it by the influence of some rare and stupendous genius, which, in a brief period, has effected more than years of patient toil had before accomplished. Some man, towering in intellect above his fellows, ordained by nature to lead into unexplored regions, and to dominate new fields of thought, has here and there made his appearance, and marked his epoch as an era. In more modern times philosophy has felt the influence of Bacon, theology that of Luther, science that of Newton, and politics that of Napoleon.

So has it been with the progress of each of the departments of the healing art. Surgery, medicine, chemistry, anatomy, physiology, and the collateral science of botany, has each in turn, since the revival of learning, felt the propulsive influence of Paré, Boerhaave, Berzelius, Morgagni, Harvey, and Linnæus.

Such an impetus was given to obstetrics late in the eighteenth century. Until that time this department was

[1] The author desires to acknowledge his obligations to Dr. S. Beach Jones, Jr., for valuable assistance in the preparation of this report.

chiefly allotted to women, and the few male practitioners who devoted themselves to it occupied a lower professional position than those engaged in medicine and surgery. He who was to establish the dignity of obstetrics, to elevate it to the position of a science, and to open the way to its rapid progress, appeared in the person of William Hunter, whose work upon the gravid uterus was published in 1774. It is true that the writings of Smellie and Levret, great contributions to obstetrics as an art, preceded it; but it is equally true that Hunter laid the corner-stone of the science in giving to the profession a work which may be said to have been to obstetrics what that of Euclid was to mathematics. It was the forerunner of the subsequent eminently valuable labours of Naeglé, and the inspiration of those of Denman and his school.

Two years before the foundation of this republic the new era of modern obstetrics was established, which has now lasted for a century. Its influence, immediately and decidedly felt in Europe, gave little evidence of its existence here, however, for the next quarter of a century, probably, in great part, for the following reasons. Its inauguration found on this continent an infant nation engaged in a struggle for independence with the formidable power of Great Britain, which taxed every resource for the following seven years: sparsely settled, without financial resources, and unprovided with the materials for sustaining a lengthy war, it became necessary for self-protection that the private resources, the individual efforts, the undivided energies of its people should be concentrated upon one single, sacred object. From this it resulted that until the year 1783 attention was entirely abstracted from the pursuits of peace—agriculture, manufacture, science, art, were all neglected.

The establishment of peace found the country entirely unprepared at once to resume those pursuits to which it had so long been a stranger. The people were impoverished, the land was unproductive, the credit of the country was not yet established, and its exchequer was empty. The immediate material wants of the inhabitants claimed their almost undivided attention, and it is not a matter of surprise that for some time we find few records of national or private efforts put forth in behalf of science in this field or in any other. As an art, practised chiefly by midwives, obstetrics was a vigorous plant, deep-rooted and strong; as a science, a delicate shoot, which feebly struggled with adverse circumstances for life; while the very seed of the sister branch of gynæcology may be said to have been unsowed.

In spite of the prejudices of the community, at an early period even in colonial times a very small number of physicians, recognizing the claims of obstetrics, devoted themselves to its practice. In 1753, according to Bartlett,[1] Dr. James Lloyd, a pupil of Smellie and Hunter, settled in Boston, and in the following year systematically began the practice of midwifery. He was the first practitioner so devoting himself of whom records can be found. In 1762 the same course was pursued by Dr. William Shippen, Jr., of Philadelphia; and these two pioneers in obstetric science began the great work here which Smellie, Hunter, and others were striving for abroad, of placing this important branch upon a level with the sister departments of medicine and surgery. The success of their efforts may be judged by the facts that in 1762 Dr. Shippen delivered a course upon obstetrics; that in 1767 Dr. J. V. B. Tennant was appointed to a special chair on this subject in New York; and that, thus introduced as a distinct department into the curriculum, the subject has to the present day been recognized as one of paramount importance and dignity.

[1] Med. Communication and Dissertations of Mass. Med. Soc., vol. ii. p. 235.

During this period essays were written by Orne, Osgood, and Holyoke[1] upon pudendal hematocele, the Sigaultian operation, rupture of the uterus, retroversion of the gravid uterus, extra-uterine pregnancy, and descriptive of cases in practice, evidencing an effort in the right direction.

In 1791[2] the operation of gastrotomy for removal of an extra-uterine fœtus was successfully performed by William Baynham upon the wife of a Virginia planter. The same gentleman[3] operated with similar success upon a negro slave in 1799. Before Baynham's first case the operation had been only once performed in this country, namely, in colonial times, by John Bard, of New York, in 1759.[4] Subsequently it was repeated by Wishart[5] and Alex. H. Stevens.[6]

The dawn of the nineteenth century found the United States ripe for progress and advancement, and while it was yet young the lamp of medical science began to burn with a brightness which it had not shown before, and which promised well for the future. We will not stop to inquire whether this improvement in progress was due to the fact that a nation, fatigued, exhausted, and impoverished by a severe conflict, had now had time for recuperation; or whether, in the language of the learned Beck,[7] this was due to the "influence which our peculiar form of government exerts over the character and progress of science." Let the sequel of this sketch prove, too, whether there be any truth

---

[1] Outline History of Gynæcology in New England, by H. R. Storer.

[2] N. Y. Med. and Philos. Journ., Jan. 1809, vol. i. p. 161.

[3] Ibid., Jan. 1809, vol. i. p. 165.

[4] Parry on Extra-Uterine Pregnancy, p. 224.

[5] Phil. Journ. of Med. and Phys. Science, 1825, N. S., vol. i. p. 129.

[6] N. Y. Journ. Med., May, 1846, p. 341.

[7] Historical Sketch of the State of Medicine in the American Colonies.

in his assertion that "it is unquestionably true that our medicine participates largely of that spirit of independence which characterizes the civil and political institutions of our country." However much patriotic zeal may prompt an inclination to accept this view; let no American ignore the fact already stated, that the way to progress in obstetrics was pointed out by the great Englishman, Hunter; and that in the very first year of the new century a similar impulse was given to gynæcology by the eminent Frenchman, Récamier.

The greatest advances which have been made in the science and art of medicine in modern times have all been due to the subordination of facts to physical investigation and demonstration. The prodigious strides made in pathological anatomy during the time intervening between Morgagni and Virchow have been due to the microscope. The great modern advance which has been made in the diagnosis, prognosis, and treatment of cardiac and pulmonary affections, has resulted from the discovery of auscultation and percussion. Diseases of the deep structures of the eye have been comprehended through the instrumentality of the ophthalmoscope. The laryngoscope has brought order out of confusion in affections of the larynx; and clinical thermometry has done more than anything else in our century to remove diagnosis and prognosis from the domain of speculation, and place them upon a scientific basis. At the commencement of the present century such an influence was evoked in behalf of gynæcology in the speculum uteri, with which Récamier, in 1801, began the study of the diseases of female sexual organs.

The history of few instruments which have come down to us from ancient times can be so clearly traced as this. Directly back through the ages which intervene between our civilization and that of the Greeks, its existence can be detected; and yet its merits and advantages had been gradually so lost sight of that Récamier may be said to have rediscovered it at the time just mentioned. The labours of this man, more than those of any other in modern times, have advanced gynæcology, and given to it its present position of dignity and usefulness.

The duty of presenting a summary of America's contribution to obstetrics and gynæcology from the time of Hunter and Récamier to our own, is truly an arduous one. So extensive is the literature which has been contributed to these subjects that even a faithful examination of it is difficult. Much more difficult is the task of separating the wheat from the chaff in the material presenting itself, for, verily, there is a surprising amount of both to be found.

In 1806, Dr. George Clark[1] reported a case of extra-uterine pregnancy, where, the head of the child presenting in the rectum, he passed his entire hand into the bowel, and, seizing the head, extracted it. Some time afterward the body and secundines were spontaneously expelled. In this case the operation proved the practicability of introducing the whole hand into the rectum without doing serious damage to that viscus, and demonstrated the utility of so doing in just such cases as that here recorded. In this course he has since been imitated by Duncan, of Edinburgh, and Jauvrin, of New York. Simon's recent advocacy of the procedure, for other purposes, is well known.

The year 1807 was signalized by two important occurrences in American medicine, the introduction of ergot into the materia medica as an oxytoxic, and the publication of the first work on midwifery which appeared in this country. Long before that time ergot had been known, and in Germany, France, and Italy had even been empirically used by midwives as a uterine stimulant. The name of the drug, indeed, in the German tongue is *mutterkorn*. To John

---

[1] Phila. Med. Museum, 1806, vol. ii. p. 292.

Stearns, first president of the N. Y. Academy of Medicine, belongs the credit of demonstrating its oxytoxic effects to the medical profession, and giving it its deserved position as the most reliable and valuable of this class of agents. His first communication, dated Jan. 25, 1807, was written from his residence in Saratoga County to Mr. S. Akerly, and published in the *N. Y. Medical Repository.*[1] This attracted great attention, gave rise to many others of similar character, and very soon obstetrics had at its disposal a most valuable agent, capable of accomplishing a result with almost certainty which none other, discovered either before or since, has been able to effect. Ergot to day stands unrivalled as an oxytoxic among drugs, and the good which has resulted from its use in post-partum hemorrhage is incalculable.

The first work upon obstetrics which appeared from the pen of a native author was that of Dr. Samuel Bard, of New York. This was published in 1807, and in 1819 had reached the fifth edition. In his preface Bard especially disclaims all originality, and declares that his work is a compend for the use of midwives and practitioners. The style of this work, though quaint, is strikingly simple, and the author appears to have been a careful, conscientious, and conservative practitioner.

"I confess," says he, "not without severe regret, that towards the end of thirty years' practice, I found much less occasion for the use of instruments than I did in the beginning; and I believe we may certainly conclude that the person who, in proportion to the extent of his practice, meets with most frequent occasion for the use of instruments, knows least of the powers of nature, and that he who boasts of his skill and success in their application is a very dangerous man."

It was during this, the first decade of the nineteenth cen-

---

[1] Vol. v., 2d Hexade, p. 308.

tury, that the greatest of all the contributions which the United States has had the good fortune to make to gynæcology, came forth from the then far west; a region from which so great an advance would at that early period have been least expected. This was the performance of ovariotomy by Ephraim McDowell, of Danville, Kentucky. The magnitude and importance of the procedure, the obscurity of its originator, and the fact that its practicability had long before been stated by eminent European authorities, all combined to render McDowell an object of distrust and obloquy. Many, both here and abroad, sympathized with the sarcastic expression of the then editor of the *Medico-Chirurgical Review:*[1] "A back settlement of America—Kentucky—has beaten the mother country, nay, Europe itself, with all the boasted surgeons thereof, in the fearful and formidable operation of gastrotomy, with extraction of diseased ovaries." Had this been stated in sober earnest, it would have been a modest and simple expression of what time, after a most searching examination, has proved to be the truth. It was, however, written in the bitterest spirit of sarcasm, the cloven foot of which is soon made apparent by the occurrence of this sentence: "Our skepticism, and we must confess it, is not yet removed."

Upon reflection, with the facts of the case clearly before us, the success of the western surgeon is not a matter of so great surprise. He was no illiterate, inexperienced, and rash adventurer, but a surgeon who had sat in his student days at the feet of John Bell and other eminent men, at that time composing the faculty of medicine at Edinburgh. "Every seminary of learning," says Sir Joshua Reynolds in one of his academical discourses, "is surrounded with an atmosphere of floating knowledge, where every mind

---

[1] Dr. James Johnson, Med.-Chir. Rev., N. S. vol. v., Oct. 1826, p. 620.

may imbibe somewhat congenial to it's own original conceptions." In Edinburgh the young American student imbibed some of this floating knowledge, and undoubtedly had the seed sown which afterwards ripened so lustily; for at that very time hints and suggestions as to ovariotomy were often thrown out by his teachers. Returning home, however, he bided his time. Before essaying his great conception he had already achieved a high reputation as a surgeon for lithotomy and hernia, and for fourteen years he cherished and reflected upon the idea of operating for extirpation of an ovarian tumour before an occasion offered for so doing.

It is evidently at variance with all the evidence at our disposal in reference to this discovery, to conclude that it was made by a sudden stroke of genius on the part of its discoverer. It should not be forgotten that what is styled genius is only the power of suddenly drawing deductions from premises slowly, carefully, patiently stowed away in the mind, studiously analyzed, and thoughtfully considered. Sir Isaac Newton expressed this opinion when, being complimented upon his genius, he replied that, "if he had made any discoveries it was owing more to patient attention than to any other talent." In our day metaphysicians are agreed in defining genius as a power of concentrating the mind in prolonged, fixed, and continued attention. Buffon tersely styles it a "protracted patience."

In 1809 the long-wished-for opportunity presented itself to McDowell,[1] and he operated successfully; then again in 1813, and again in 1816; although he did not publish these cases till 1817. What a commentary upon the grand nature of the man was this calm deliberation and hesitancy to rush into print! He had performed an operation never before attempted in the history of the world, and with three suc-

---

[1] Eclectic Repository and Analytic. Rev., April, 1817, p. 242.

cessive good results, and yet he did not hasten to blazon it abroad!

A great deal has been said, and very properly said, concerning the fact that McDowell got the suggestion of ovariotomy from abroad, and only developed it afterwards in his own country. Even had McDowell never lived, America seemed destined to be connected with this great surgical triumph from its inauguration; for in July, 1821, Nathan Smith, then Professor of Surgery in Yale College, performed ovariotomy[1] entirely without the knowledge of the fact that he had been preceded by one of his own countrymen in 1809, and by a German in 1819.

The scope of this paper will not admit of a record of the names of the immediate followers of these surgeons; suffice it to say that before the year 1850 eighteen operators had successfully performed thirty-six operations, with twenty-one recoveries and fifteen deaths.

In England ovariotomy was never performed till 1836; in Germany it was first performed in 1819, and in France in 1844.

It will thus be seen that this operation, remarkable at once for its simplicity and efficiency, did not rapidly advance to a recognized place as one of the resources of surgery, but slowly and painfully overcame the prejudices and doubts of worthy men, and the misrepresentations of detractors. In effecting this result, America by no means stood alone. Nevertheless it was to Americans, the successors of McDowell, that it was in great part due. The names most intimately connected with the work are those of John L. and W. L. Atlee, Dunlap, Peaslee, and Kimball. To the Atlees too much credit in this regard cannot be accorded. Profoundly impressed with the importance and future usefulness of the procedure, they pressed onward in the work of

---

[1] Am. Med. Recorder, 1822, vol. v. p. 124.

establishing its claims with that dignified indifference to the criticisms of opponents which always characterizes successful innovators. They operated upon all suitable cases, when each venture insured a storm of censure; when every fatal result was cited in evidence of their recklessness; when persistence robbed them of the esteem of many whose good feeling they could not but value.

"On the 17th September, 1843," writes Alexander Dunlap, of Springfield, Ohio, "I performed my first ovariotomy, and carefully wrote out the case for publication, and sent it to a medical journal. They sent it back, with a note, stating that they could not publish the case of such an unjustifiable operation. I threw it into the waste-basket, determined to write no more for medical journals; but, being satisfied that I was right, to continue the operation. From that time, for a number of years, I was looked upon by most of the profession out west as a kind of an Ishmaelite in the regular profession in regard to surgery, and in that operation in particular. . . . I have now operated 106 times for ovariotomy (1876), with 27 deaths and 79 cures." Peaslee operated first in 1850 and Kimball in 1855.

Let it be borne in mind that these operators, with a few others, for a long time, stood almost alone. In those days it was as difficult to find a physician bold enough to sustain the operation as it now is to find one who dares decry it, and the wisest and most eminent surgeons of our country did not hesitate to declare,[1] "that in a few years the measure will be consigned to the oblivion it so richly merits." This, indeed, was a mild expression of disapprobation compared with many others from the best men in our ranks. Let us rather draw the veil over the exhibition of vituperation and personal abuse which disgraced the opposition, and strive to forget that the bigotry and narrow-mindedness

---

[1] Liston and Mütter's Surgery, 1847, p. 422.

which endeavoured to crush the great discovery of Copernicus still lived in our day, to strive against ovariotomy. "Pride," says Sir William Hamilton, "has led men to close their eyes against the most evident truths which were not in harmony with their adopted opinions. It is said that there was not a physician in Europe above the age of forty who would admit Harvey's discovery of the circulation of the blood."

To-day, when ovariotomy is generally accepted as a valuable surgical resource, it is difficult for one to appreciate the reasons for the tardiness with which it overcame European prejudices, and forced its claims upon the notice and confidence of men who have since learned to accord to the procedure its true value. So entirely has this disposition on the part of trans-Atlantic surgeons been now overcome that a very general and, we are forced to say, a very reasonable feeling of surprise has been excited in America at what has seemed to be an inclination to ignore her indisputable rights in the matter.

"Till 1858," says an ovariotomist of Great Britain, as well known for his personal excellence as for his skill and success as an operator, writing as late as 1873, "I could find nothing whatever anywhere to encourage, but everything to deter one from attempting it. Ovariotomy was then, as an operation, simply nowhere."[1] This was a mistake. Ovariotomy since 1809 was somewhere; namely, in the land in which McDowell had performed thirteen operations with eight undoubted successes before 1830; W. L. Atlee[2] fourteen operations prior to 1851; J. L. Atlee[3] double ovariotomy in 1843; and where over twenty-five other surgeons had removed ovarian tumours each one or

---

[1] Mr. Thos. Keith, Brit. Med. Journ., Dec. 20, p. 739.
[2] Am. Journ. Med. Sci., N. S., vol. xxix. p. 387.
[3] Ibid., N. S., vol. vii. p. 44.

more times prior to the year in which this eminent commentator began to discover the whereabouts of the procedure. The only extraordinary thing connected with the matter is that so important an operation could for almost an entire half century have so completely concealed its huge proportions from the ken of so acute an observer, and that too in a country teeming with medical periodicals and a nation not prone to hide its light under a bushel.

England, France, and Germany have each in turn been claimants of an operation, which after the most critical and thorough search stands fully accredited to America. "In faith, 'twas strange, 'twas passing strange," and yet 'twas true, that a surgeon of the Western wilds, with what Piorry once styled "une audace Américaine," stole a march upon the polished savans of the old world, as if in the silent watches of the night.

It is difficult to estimate the amount of good which this operation has bestowed upon humanity! Practised to-day in every civilized country in the world, yielding the statistics of seventy to seventy-five per cent. of recoveries, and daily being improved in its various steps, it may well be regarded as one of the greatest surgical triumphs of the century. "It may be shown," says Peaslee, "that in the United States and Great Britain alone, ovariotomy has, within the last thirty years, directly contributed more than thirty thousand years of active life to woman, all of which would have been lost had ovariotomy never been performed." To have done this even for one generation alone is glory enough for one mortal, and his country, apparently in recognition of this fact, leaves his grave without a mark, and his memory to be preserved only in the hearts of the thousands of grateful women whom his genius has saved from death.

Should the day ever arrive in which the memory of McDowell shall be honoured by a monument, surely no one will deny to it the right to that inscription which declares upon the statue of Washington; he "has rendered his name dear to his fellow-citizens, and given the world an immortal example of true glory."

In the year 1816,[1] John King, of Edisto Island, South Carolina, performed one of the most remarkable operations for removal of an extra-uterine fœtus ever placed on record. The case was one of abdominal pregnancy; the head presented in the pelvis, outside of the vagina; he cut through the walls of the latter, and applying the forceps, while abdominal pressure was exerted upon the child from above, had the rare good fortune to save both mother and child.

Towards the close of the eighteenth century there arose a man whose genius left its impress upon American obstetrics more decidedly than that of any other has done before or since. Decided in opinion, vigorous in expression, terse in argument, and trenchant in style, he did a great deal towards elevating the department to which he devoted himself. William Potts Dewees was born in 1768 and died in 1841, after a long and laborious professional career, during which he exerted a powerful influence as Professor of Midwifery in the University of Pennsylvania, and a writer upon obstetrics, gynæcology, and pediatrics.

In after times it is impossible to estimate the degree of influence which has been exerted by such a man as Dewees. It is a matter of tradition only, and we can merely point to the *literæ scriptæ* which outlive him. He contributed a Treatise on the Diseases of Females (1826), which went to the tenth edition; a Treatise on the Physical and Medical Treatment of Children (1825), which reached the tenth edi-

---

[1] Med. Repository, 1817, N. S., vol. iii. p. 388. See also "An analysis of the subject of extra-uterine fœtation, and of the retroversion of the gravid uterus, by John King, Esq., of South Carolina." Norwich, 1818, 8vo. pp. 176.

tion; and a Comprehensive System of Midwifery (1824), which went into a twelfth edition. Of the last Prof. Hugh L. Hodge[1] declared "it takes a stand decidedly in advance of Denman, Osborne, Burns, and other English authorities in general use in our country at that period, and even of Baudelocque himself in throwing aside from his excellent system much that was useless, and, it may be said, imaginative."

Dewees had two able successors in Meigs and Hodge, both of whom reached old age in the active performance of their professional duties, and left indelible traces of their influence by reason of their strong intellectual qualifications, valuable literary contributions, and rare personal worth.

Charles D. Meigs was born in Bermuda in 1792, and commenced practice in 1815. For many years he filled the chair of obstetrics in the Jefferson Medical College of Philadelphia, and contributed largely to medical literature. His most important works were Woman, her Diseases and Remedies (1847); Obstetrics, the Science and Art (1849); a Treatise on Acute and Chronic Diseases of the Neck of the Uterus (1850); and on the Nature, Signs, and Treatment of Childbed Fevers (1854).

Meigs[2] drew special attention to cardiac thrombosis as a cause of those sudden deaths which occur in childbed, and which had generally been attributed to syncope. "I had noticed, on various occasions, the total want of any means of explaining such disasters," says he, referring to sudden deaths post partum, "and remained as much in the dark as my compeers, until I discovered that the incident depends most commonly on the sudden coagulation of the blood that occupies, for the time, the right auricle of the heart, and, in

---

[1] Memoir in Amer. Journ. of Med. Sci., Jan. 1843.
[2] Med. Examiner, March, 1849, p. 141.

some of the cases, even that which is in the ventricle, and the pulmonary artery."[1]

It has been remarked by an eminent American author that Meigs "just escaped the honour, which is now, and will hereafter be given to the eminent Virchow, of Berlin, of a great pathological discovery." Even admitting the truth of this statement, it is certainly well that the justice of the award should here be questioned. Meigs proclaimed the fact in no uncertain or wavering tones, but boldly, decidedly, repeatedly, and by every method. Why is the honour not his? What else could he have done to deserve it? Many of his countrymen will sympathize with the voice which speaks now, after death, in this unmistakable manner. "I have a just right to claim the merit of being the first writer to call the attention of the medical profession to these sudden concretions of these concrescible elements of the blood in the heart and great vessels." It may be said that he did not follow his discovery into detail as regarded secondary deposits of emboli. What of that? He does not claim to have done so. What he does claim is clearly and unquestionably claimed with justice.

The style of Meigs was peculiarly quaint and antique. Yet he possessed in a remarkable degree the power of fixing salient points upon the mind of the reader or listener, and burning into the memory the maxims which he deemed of greatest importance. Meigs died June 22, 1869.

Hugh L. Hodge practised in Philadelphia from 1818 to 1873, during which period he exerted a wide and decided influence as Professor of Obstetrics and Diseases of Women in the University of Pennsylvania, and as the author of a number of valuable works upon these and kindred subjects. His most valuable contributions to literature were Cases and Observations regarding Puerperal Fever (1833); Dis-

---

[1] Treatise on Obstetrics, 5th ed., p. 352.

eases Peculiar to Women, including Displacement of the Uterus (1860); Principles and Practice of Obstetrics (1864); and Essays upon Synclitism of the Fœtal Head (1870–71).[1] In his essays and lectures Hodge made prominent, by precept and illustration, the value of forceps as compressors in ordinary delivery, and after perforation; synclitism of the fœtal head; the importance of the induction of premature labour where even without pelvic deformity repeated fœtal deaths have occurred from premature ossification; the prophylactic influence of mechanical support in prevention of habitual abortion, and its efficacy in cases of uterine fibroid; and added to the *armamentarium obstetricum* "Hodge's Forceps," the instrument more generally used in this country than any other; a compressor cranii; a craniotomy scissors; and placental forceps—all attesting rare mechanical ingenuity.

For gynæcology Hodge accomplished much by the origination and development of two ideas which have already done a great deal of good, and will in the future do more than they have yet accomplished. The first of these involved the recognition of the fact that that state of the uterus characterized by enlargement, tenderness, displacement, congestion, and hypersecretion is not "inflammation," and should not be treated as such; the second, that a double vaginal lever can supplement the exhausted uterine supports under these circumstances, and by sustaining the uterus give great relief to all these conditions. With regard to the first of these views, nothing more can be stated here than the expression of the belief that it constitutes one of the most important facts in uterine pathology. As to the second, something is necessary. Prior to the time when Hodge began a course of careful, laborious, and conscientious experiments upon the shape, material, uses, and varieties of

---

[1] Am. Journ. Med. Sci., Oct. 1870, p. 325, and July, 1871, p. 17.

pessaries (1830), these instruments had been used both in this country and in Europe. Indeed, even as long ago as the period of the Greek civilization they had been employed. But the disk, the globe, and similar instruments were badly contrived, did not depend upon any true mechanical principle, and accomplished little by comparison with what Hodge's improved instruments have since done. He introduced the philosophical double lever, gave accurate and precise rules for replacing and sustaining the displaced uterus by it, insisted upon every pessary sustaining itself against the vagina instead of against the rami of the pubes, particularly urged that after being placed both uterus and pessary should be movable in the pelvis, and thus brought a subject which had before belonged to the realms of empiricism into the precincts of science. Hodge's pessary is unquestionably the parent stem from which the host of excellent modifications now existing took their rise. It may safely be averred that he accomplished more for mechanical support of the uterus than any one has ever done before or since his time. He first constructed these instruments out of silver which was plated lightly with gold, but in time he used vulcanite or hard rubber entirely.

Hodge once gave to a friend this account of the consummation of the discovery of the lever pessary: "He had been contemplating for a long time the subject of new shapes for pessaries, and after many experiments had found nothing satisfactory. One evening, while sitting alone in the room where the meetings of the Medical Faculty of the University were held, his eyes rested on the upright steel support by the fireplace, designed to hold the shovel and tongs. The shovel and tongs were kept in position by a steel hook, and as he surveyed the supporting curve of this hook, the longed-for illumination came; the shape, apparently so paradoxical, revealed itself in the glowing light and flickering flame of the burning grate, and the Hodge lever pessary was the result."[1] A sudden effort of genius, was it? No: this was the moment at which the detached thoughts, long and carefully stored away in the inventor's mind, combined to form a harmonious whole. The steel hook did for his mind what the swinging church lamps did for that of Galileo in suggesting the pendulum.

Henry Miller was born in Kentucky, in 1800, commenced practice in 1821, and published a work upon Human Parturition in 1849, and upon the Principles and Practice of Obstetrics in 1858. He was for a long period Professor of Obstetrics and Diseases of Women in the University of Louisville, and both as teacher and writer decidedly influenced the department to which he devoted his energies. We owe to him the method of making the application of fluid caustics to the cavity of the body of the uterus by saturating with them a cotton wrapped rod or probe, and he was among the first to adopt the use of the speculum uteri in the great West, and the first to employ in that part of our country anæsthesia in midwifery.

In 1819 the chair of obstetrics in the College of Physicians and Surgeons of New York was filled by one who will long be remembered for his eloquence, erudition, and rare geniality—John W. Francis. Unfortunately for the department which now engages our attention, Dr. Francis turned his literary efforts in the direction of general medicine, literature, and pathology. Little remains to us of his obstetrical writings except his copious annotations of Denman's Midwifery, which he edited in 1821. A perusal of these makes one regret that he did not leave behind him more extensive contributions to this department embodying more of his large experience and acute observation.

The next systematic writer upon this department in the

---

[1] Discourse Commemorative of H. L. Hodge, M.D., by R. A. F. Penrose, M.D., Phila., 1873.

United States was Gunning S. Bedford, who practised in New York from 1830 to 1868; was for over twenty years Professor of Obstetrics in the University Medical College, and published a work on Diseases of Women and Children in 1855, and another on the Principles and Practice of Obstetrics in 1861.

In the year 1841, Bedford established, in connection with the University Medical College of New York, the first clinic for the diseases peculiar to women ever held in this country. This he maintained with great ability, energy, and enthusiasm, and from it he gleaned the material for a work which created a very decided sensation both in this country and in Europe. This clinic, under the care of his able successors, Charles A. Budd and M. A. Pallen, still exists. But it has been the parent of many other similar ones not only in New York but throughout our country. No medical school, indeed, is now considered complete without such a sphere for the instruction of students, and a vast deal of good has resulted from his move in this direction.

Thus far this essay has chiefly dealt with the labours of those of a past generation who, in the early part of the century, sustained the department of which we write; and the careers as well as the works of individuals have been noted. From this point we are called upon to undertake the more delicate and far more difficult task of dealing with the labours of our contemporaries. The chief reason for the difficulty and delicacy of this duty grows out of the fact that nothing is harder than to arrive at a just appreciation of the merits of contemporaries, more especially on the part of those labouring in the same field. Prejudice, personal bias, and that tendency which all men feel to undervalue what is at their own doors, and exalt that to which distance lends enchantment, all combine to defeat a just, fair, and generous estimate. Then, too, the umpire, however conscientious and unprejudiced he may be, lacks the great assistance which the test of experience alone can give in deciding as to the value of new procedures and the credit which should be accorded to their discoverers. He has not the opportunity of learning the verdict of time as to what is and who are the fittest to survive; of that "wise beneficent law by which the improvement and perfection of the human race alone can be secured; that law in consequence of which the best specimens of a species survive, and become the progenitors of generations more perfect than those preceding them." The only feelings which can sustain him who makes the effort and render him impervious to the shafts of criticism, is an abiding faith in the rectitude of his intentions, and in the sincerity of his efforts to render to every man, without prejudice or favour, what he honestly regards as his just dues.

The plan which suggests itself as best is to notice, 1st, the original discoveries which have proved of greatest practical value; 2d, the most striking and important contributions to periodical literature and systematic works upon the subject; 3d, instruments and mechanical contrivances of greatest importance.

In the year 1841 a most important contribution was made to the treatment of peritonitis by Alonzo Clark,[1] of New York, in the introduction of the plan now known as the "opium treatment." In the spring of that year Dr. Clark saw several cases of this disease treated by Armstrong's method—a full bleeding, and a full dose of opium, to prolong the effects of the bleeding. He was impressed with the idea that opium was the curative agent, not the bleeding. In the next three years he treated all the cases he met on that idea, giving opium, or an opiate, in full and frequent doses, and nothing else. The result was that just

[1] New York Journ. of Med., Jan. 1858, p. 82; and Ramsbotham's Syst. of Obstetrics, Am. ed., Phila., 1855, p. 533.

eight out of nine cases were cured. A success very encouraging, but not quite so marked, attended the similar, subsequent use of the drug. With this experience he resolved to give it a trial in puerperal peritonitis. The opportunity, however, did not occur till 1848. The first trial was successful, the patient taking 100 grs. of opium in four days. Between that date and 1852 nothing occurred that was decisive regarding the merits of the plan. But in the latter year an epidemic of puerperal fever occurred at Bellevue Hospital, in which the exclusive opium treatment was fully tested by him. In the first case it failed, or rather through the timidity of the House Physician, it was not tried—only three grains of opium were given in twenty-four hours. A few days later four cases came under his care at once. He assigned them to another member of the House Staff, a man of more decided character, now a distinguished surgeon and sanitarian, with detailed instructions. All of these were cured. It was in the course of this outbreak that the opium treatment for puerperal peritonitis was shown to be the best that had then, or has since, been proposed. This physician assisted in the treatment of puerperal fever in the same hospital twelve years earlier, in which, out of thirty attacked, twenty-nine died. Nothing was then known of the antiphlogistic power of opium.

The quantity of opium, or one of its alkaloids, required to subdue and control the inflammatory process varies greatly. In some cases two grains of powdered opium every two hours answers the purpose, while in others eighty drops of Magendie's solution (xvj grs. of morphia to one ounce of water) every two hours for six or seven doses are required. Dr. Clark records a case in which the patient took during "the first twenty-six hours, of opium and sulphate of morphia, a quantity equivalent to 106 grains of opium; in the second twenty-four hours she took 472 grains, on the third day, 236 grains, on the fourth day, 120 grains, on the fifth day, 54 grains, on the sixth day, 22 grains, and on the seventh, 8 grains."

By this system a tolerance of the drug is rapidly effected, pain is annihilated, nervous and mental disquietude relieved, and the most satisfactory results commonly attained. While it is put in practice, however, a physician should constantly remain by the bedside to detect the development of dangerous narcotism, and combat it by appropriate means. It is surely not claiming too much for Clark's method to assert that it surpasses in efficacy all others which have yet been made known to the profession.

In 1844, Dr. J. C. Nott,[1] of Mobile, published a case of "coccygeal neuralgia," in which he practised extirpation of the bone, which proved to be carious, with entire relief to his patient. This was the first time that either this disease or its remedy had been described. At a later period Sir James Simpson, not knowing of Nott's essay, described the disease under the name of coccyodynia or coccygodynia, and advocated the same method of treatment.

Although a decided impulse was given to gynæcology by the introduction of the speculum by Récamier, a great need was felt of something which would expose the uterus and vagina to more complete and satisfactory investigation. For want of this the cure of vaginal fistulæ had thus far proved impracticable, and many operations upon the uterus itself difficult of accomplishment. In 1852 there appeared an article from the pen of a hitherto unknown author, which changed all this, and threw a flood of light into dark places. This was an essay upon vesico-vaginal fistula, by J. Marion Sims,[2] then of Montgomery, Alabama, in which he introduced a speculum which developed a new principle of ex-

---

[1] New Orleans Medical Journal, May, 1844, p. 58; and Am. Journ. Med. Sci., Oct. 1844, p. 544.

[2] American Journ. Med. Sciences, Jan. 1852, p. 59.

amination of the uterus and vagina. The discovery of a method of cure for vaginal fistulæ was a great stride onward, but the method of examination by retraction of the perineum and posterior vaginal wall, while the body of the patient is so placed as to secure distention of the vagina by air, has served to give to gynæcology an impulse second in importance only to that given by Récamier. Récamier's discovery lifted this department from the field of speculation to that of science. Sims has served to advance it very greatly beyond the point which it would have occupied if reliance were still placed upon previous methods.

Important discoveries are not made suddenly as if by one leap on the part of some great intellect. They are arrived at slowly, step by step, and by the workings of many minds; as many unseen influences slowly mature a harvest which in due time falls to one sickle. The inspiration of discoverers is the offspring of the times in which they live; such men are exponents of the mental workings of their period, mouth-pieces of the civilization which developed them. The resultant of the premises evolved from ten great minds of one decade are often combined in the deductions of a single genius in another. Hence it is that discoveries are often simultaneously made in various parts of the world by men who have had no communication with each other, and that their origination is invariably disputed by rival claimants. Morse discovered the telegraph, but ever since Franklin's kite brought down the lightning from the skies, many others had been preparing the way for him. Wells discovered anæsthesia, but for many years before, school-boys had for their amusement been painlessly bruising themselves under the influence of laughing-gas, never dreaming that the means of securing unconsciousness of pain which they adopted would one day become systematized and utilized as a great boon to humanity.

These remarks find no more perfect illustration in the

discoveries of surgery than that of the cure of vesico-vaginal fistula. The writings of the Greek, Roman, and Arabian schools of medicine are singularly silent with reference to an accident which has a striking faculty of pressing itself upon the attention, and must have been very common before the days of the Chamberlaynes. From the times of Paré, however, it attracted the special care of surgeons, and year after year efforts were made to close these small, but important, lesions. It would take too much space to tell of the efforts of Paré, Roonhuysen, Vœlter, Fatio, and many others; suffice it that at the beginning of the nineteenth century nothing had been accomplished. In the eighteenth, however, "coming events cast their shadow before," and the glimmer of the dawn became visible in the operations, and occasional successes of Desault, Naeglé, Schreger, Lallemand, and Roux. In 1834 Gossett, of London, absolutely discovered the method of cure, and, his labours being forgotten, Metzler, of Prague, in 1846, again did so. And now, too, Hayward and Mettauer, of this country, began to get good results. But sporadic, desultory, haphazard results mark a different era from systematic and certain ones, and the matter may be said to have been really little advanced till Marion Sims published to the world his method of treating these accidents, which was at once so simple and systematic as to place the procedure at the disposal of every surgeon.

No more forcible comment can be made upon the perfection of Sims' procedure than the mere citation of the fact that even now it stands, for the great majority of surgeons, virtually unaltered, and as simple in details as when it left the master-hand.

Various modifications have been suggested both in this country and in Europe. Chief among these are the clamps of Battey, Atlee, and Bozeman. The last of these only deserves special mention on account of the excellent results which have been obtained with it by its originator, Nathan Bozeman. This operator, who was the earliest to follow Sims in this field of surgery, and who has devoted himself to it with an earnestness which has been surpassed by that of no other, has always preferred a modification of the knee-elbow position to that on the side, and has approximated the pared edges of the fistulous orifice by passing his sutures through a leaden shield, or, as he styles it, a "button suture." By this method very gratifying results are obtained, and after an experience of more than twenty years with it, its originator still employs it with confidence in its advantages over the suture alone.

The medical profession in New York, recognizing the value of Sims' discoveries, warmly endorsed an effort on his part to establish a Woman's Hospital in that city, where, thanks to the well-known generosity of its citizens, such an institution was founded in 1855. From this institution, through the labours of Sims and his able coadjutor Thomas Addis Emmet, a great deal has emanated for the advancement of gynæcology. To these two men a great deal of credit is due for establishing and disseminating an exact and systematic method in the study of the diseases of women. The greater facility afforded for operations upon the vagina and uterus by Sims' method of examination, has accomplished an improvement in all such procedures, and these two operators, who were first in the field with this advantage on their side, have been greatly instrumental in this result. Operations upon the perineum, upon fistulæ, upon constricted and tortuous uterine necks, upon voluminous and atonic vaginæ, have all felt this influence. Posterior instead of lateral section of the cervix for anteflexions of body and neck, is a good example of such an improvement as has been thus effected.

Until the establishment of Marion Sims as a specialist in diseases of women in New York about the year 1852, no one in this country had heretofore devoted himself to this department to the exclusion of general practice. By him and by T. A. Emmet and H. R. Storer more than any others, this practice was established. That a great deal of good has resulted from the devotion of able minds to the special investigation of this subject, no candid observer can doubt. And yet every thoughtful man who wishes well to the department, must view with concern the unwise haste with which young practitioners, who have had neither time nor opportunity to acquire experience in general medicine, strive to devote themselves to it. Can it ever be that he who knows little of the management of the diseases which affect the peritoneum, stomach, lungs, and liver, can deal efficiently with the disorders of an organ or set of organs which are especially affiliated with them in all their variations of disease, in all their physiological functions? He who deals efficiently with the whole, may in detail deal with all its parts, but he who learns to deal with a part alone, can never be equal to coping with the whole.

Before the introduction of Sims' method of uterine examination, the use of the vaginal tampon, the most important of all hemostatic means in connection with the non-pregnant uterus and with this organ up to the fifth month of pregnancy, was difficult, painful, and unreliable. The introduction of a silk handkerchief, a kite-tail tampon, a mass of cotton, a muslin bandage, and all similar materials, was very unsatisfactory. The most perfect facility and efficiency attend tamponing the vagina with wet cotton while the patient lies upon the side, and the vagina is dilated by means of the duck-bill speculum. Pieces of cotton soaked in water, pressed and flattened out by the fingers, each about the size of a very small biscuit, are pressed into the vaginal cul-de-sac by means of forceps till this is filled. Then other pieces are packed firmly around the cervix until only the os is visible—a smaller pad is then pressed firmly against or introduced within the cervical canal, and the whole vagina is then filled to its lowest portion.

At a meeting of the American Medical Association in 1853, a prize was awarded to a very remarkable and valuable essay by Dr. W. L. Atlee,[1] entitled, "The Surgical Treatment of Certain Fibrous Tumours of the Uterus heretofore considered beyond the resources of Art." In this Dr. Atlee advocated the removal, by enucleation, of tumours which up to this time were looked upon as incurable, and by his brilliant results he led the way to a plan of treatment which has been productive of a great deal of good. His plan of treating these growths is now very commonly adopted by practitioners who appear to forget to whom the heroic and life-saving method is due. Even as early as 1850, Prof. Mussey of Ohio remarked, "Of all the achievements of modern surgery, we meet with none more striking or extraordinary than the operations performed by Professor Atlee for the removal of intra-uterine fibrous tumours."

In 1854, a gold medal was awarded by the Ohio State Medical Society to Dr. M. B. Wright of Cincinnati, for an essay entitled "Difficult Labours and their Treatment." In this essay the operation of bimanual version was so fully, so clearly, so unmistakably described, that it is difficult to understand how many of his countrymen could have since permitted themselves to style the procedure by any other name than "Wright's Method." An examination of the written testimony bearing upon the subject, certainly seems to give endorsement to the following claim on the part of Wright.

"I claim the credit, if credit there be, of having first suggested to the profession, and demonstrated in practice, the value of bimanual version."[2]

It must be understood that Wright neither claims nor

---

[1] Trans. Am. Med. Association, 1853, vol. vi. p. 547.
[2] Letter to the author of this review in Jan. 1876.

deserves the credit of the discovery of bimanual version as a procedure, but only that of the method of its performance. Flamand long before him described cephalic version by this method, but Wright improved upon and simplified the procedure. This is Wright's description of his plan:—[1]

"Suppose the patient to have been placed upon her back, across the bed, and with her hips near its edge—the presentation to be the right shoulder, with the head in the left iliac fossa—the right hand to have been introduced into the vagina, and the arm, if prolapsed, having been placed, as near as may be, in its original position across the breast. We now apply our fingers upon the top of a shoulder, and our thumb in the opposite axilla, or on such part as will give us command of the chest, and enable us to apply a degree of lateral force. Our left hand is also applied to the abdomen of the patient, over the breech of the fœtus. Lateral pressure is made upon the shoulders in such a way as to give to the body of the fœtus a curvilinear movement. At the same time, the left hand, applied as above, makes pressure so as to dislodge the breech, as it were, and move it towards the centre of the uterine cavity."

All controversial topics should be avoided in an essay like the present, but it would be discourteous to a distinguished English obstetrician not to note the fact that he has doubted the claim of Wright to originality in this matter. In a letter published in the *Amer. Journ. of Obstetrics*, etc., for Feb. 1873, Dr. Braxton Hicks, of London, says:—

"Now the distinctive point of the plan I have introduced was just this, that *both hands are used together*, one supplementing the other, so that when the internal hand began to lose power the external one would begin to gain power, and *vice versa*. This principle was applied by me to both partial and complete version, and it is (as far as I have been able to discover) a curious fact that in the practice of neither German nor other obstetricians has the use of the two hands simultaneously been de-

---

[1] Trans. Ohio State Med. Soc., 1854, p. 82.

scribed. The only use of the outside hand has been hitherto to steady the uterus to prevent recession. This character it is which Dr. Richardson[1] has overlooked, and it is for this that I am desirous of claiming for *myself* whatever of originality it possesses."

This claim is perfectly clear, and can be answered without difficulty or circumlocution. Wright says "at the same time the left[2] hand, applied as above, makes pressure so as to dislodge the breech, as it were, and move it towards the centre of the uterine cavity." Surely no one can suppose that this means that the left hand merely steadies the uterus. Cazeaux declares that Flamand got hold of the head with the hand in the vagina, "if the efforts made by the other hand through the abdominal walls, have not proved sufficient to make it descend into the excavation."

There is no question as to the fact that Dr. Hicks has done a great deal of good in simplifying podalic version by this method. But the extension and utilization of a method is not here at issue; it is the origination of the principle which is in question.

Even had Wright not made this advance, it seemed destined to be made in America, for in the next year Penrose,[3] of Philadelphia, in an article entitled "Cephalic version in shoulder presentations, with the arm in the vagina," described bimanual version without a knowledge of the fact that he had been anticipated by Wright.

During the course of the same year a very valuable contribution was made to the treatment of septicæmia following ovariotomy, by E. R. Peaslee,[4] of New York. His method was the introduction of a catheter or similar tube into the

---

[1] Who maintains Wright's claim.
[2] The right hand is in the vagina.
[3] Medical Examiner, July, 1855, p. 405.
[4] Am. Journ. Med. Sci., April, 1863, p. 355, and July, 1864, p. 47. See also Amer. Journ. Obstet., 1870, vol. iii. p. 300.

peritoneal cavity and boldly washing out this serous sac, interference with which had for all time been regarded with so much dread. Experience with the plan, extended now over a period of twenty years, stamps it as a reliable method of meeting one of the most dangerous consequences of this grave operation, and corroborates the high estimate which was put upon it in the early days of its existence. Unquestionably many lives have been saved by a timely resort to it. In one of Peaslee's early cases the use of intraperitoneal injections was kept up for fifty-nine days, and in another for seventy-eight days. In both of these cases recovery took place as a reward for the prolonged and persevering efforts of the fearless innovator.

In 1856, Sims made known his operation for narrowing the vagina for the cure of prolapsus uteri. In this he had been anticipated by Dieffenbach, Heming, and other Europeans, but his method was an improvement over others, and was a revival of what had fallen into almost entire disuse.

In the same year,[1] Dr. James T. White, of Buffalo, reduced by taxis an inversion of the uterus of eight days' standing. In his report of this case he took occasion to predict that the profession would soon alter its views with regard to the practicability of reposition in chronic cases, a prophecy which was happily fulfilled, in great degree in consequence of his own labours, two years afterwards.

Daillez,[2] who published a thesis upon this subject as early as 1803, reported a case of reduction by taxis as late as the eighth month after occurrence of the accident; another is reported in 1847; and even as late as 1852, Canney and Barrier are declared to have accomplished it. But the plan was not systematized and placed upon the basis of a recognized and legitimate procedure until 1858, when White of

---

[1] Buffalo Med. Journ., vol. xi. p. 596.
[2] Colombat, Dis. of Women, Am. ed., p. 186.

Buffalo, and Tyler Smith of London, simultaneously replaced uteri in the condition of chronic inversion, and gave to the procedure the position of a standard operation.

Up to the present date White has successfully reduced by taxis twelve cases, extending from seven months to twenty-two years in duration.

In 1858, Gaillard Thomas[1] published an essay upon the treatment of prolapse of the funis by gravitation developed as a remedial measure by placing the patient in the genu-pectoral position. This plan, which it appears had been formerly in use, had been so entirely lost sight of, that for ten years after its introduction by him, the fact of its previous existence was not known. Since the time of his article it has come into general use as the most rational and simple method of treating this accident during the earlier stages of labour.

The intractable nature of, and extreme distress attendant upon chronic cystitis, are too well known to require mention. For a long time the attention of American surgeons has been directed to the relief of this condition by surgical means. In 1846,[2] Willard Parker created a recto-vaginal fistula in the male for the removal of a stone, and being struck by the relief afforded to a cystitis which existed, he subsequently repeated the operation for the relief of the latter condition in men between that time and 1867, when he read an essay upon the subject before the New York State Medical Society. "The object in view," says he, "was to open a channel by which the urine could drain off as fast as secreted, and thus afford rest to the bladder, the first essential indication in the treatment of inflammation." In 1867, Paul F. Eve followed Parker's example in thus operating upon the male. But in

---

[1] Trans. of N. Y. Acad. of Medicine, vol. ii. p. 21.
[2] Transact. of N. Y. State Med. Soc., 1867, p. 345.

1861,[1] Nathan Bozeman applied the procedure to the female bladder with the result of curing chronic cystitis

Without a knowledge of any of these facts the same idea suggested itself to the minds of Sims and Emmet[2] as early as 1858, and at a later period, 1861, the latter of these gentlemen, at the suggestion of the former made three years before, practised the operation for chronic cystitis in the female. Although the origination of the method does not belong to Emmet, to him is justly due the credit of having systematized the procedure, and placed it upon the basis of a recognized surgical resource. Whether it is destined to give way before the less serious procedure of distending the urethra, and thus establishing incontinence of urine, time will prove. That it is in itself a most valuable operation, no one can doubt who has seen the relief afforded by it to women nearly exhausted by ceaseless vesical tenesmus, loss of sleep, and nervousness.

In Smellie's[3] *Collection of Preternatural Cases and Observations in Midwifery*, vol. iii. p. 232, will be found evidence of the fact that that great obstetrician recognized the value of gravitation, developed by placing the patient in the genu-pectoral position, as an aid to the operation of podalic version. He mentions his having repeatedly resorted to this posture in performing version, but does not claim originality for it, as he styles it "Daventer's method." The first case in which Smellie resorted to it occurred in 1753. In Wright's pamphlet, already alluded to, published in 1854, and entitled "Difficult Labors and their Treatment," the following passage occurs on page 23: "The hand can be more readily introduced into the uterus, and the feet reached, however, with the patient on her elbows and knees, than when on the back

---

[1] Transact. of N. Y. State Med. Soc., 1871, p. 326.
[2] Amer. Practitioner, Feb. 1872, p. 65.
[3] Ibid., Jan. 1876, p. 59.

or sides. There may be cases, in which advantage would be gained, by placing the patient in this position, preparatory to cephalic version."

It will be observed that Smellie resorted to the knee-elbow position as an adjuvant to podalic version, and that Wright very cautiously offers it as a mere suggestion. To P. R. Maxon, of Syracuse, N. Y., belongs the credit of having established the claims of this method in the performance of cephalic version in cases of transverse presentation. He thus describes the procedure in the case of a lady who had previously lost three children by podalic version.

"Remembering the fate of the other children, and finding this one very large, I suggested the feasibility of correcting this shoulder presentation in the same manner as I had corrected the abdominal in the first instance. With his (the attending physician's) consent, I made the effort in the following manner: I folded several quilts compactly, laying them upon one another to the height of about one foot, and assisted her to kneel upon the quilts with her head and shoulders resting upon the bed and her face forwards, so as to bring her body to an angle with the bed of nearly 90°. I then pressed my hand gently against the shoulders, which readily receded, until I was enabled to clasp the vertex with my fingers, and with the assistance of the next pain to so 'engage' it that, when the patient was placed upon her left side and the quilts removed, a perfectly natural presentation presented itself. In a few hours the labour terminated in the delivery of a healthy boy, weighing ten pounds."

No one who has not resorted to Maxon's method can appreciate the great facility with which a shoulder or even an arm presentation may be altered into one of the vertex; and no one who has done so will doubt the great value of the plan. Of course, after the amniotic fluid has been long evacuated, and the uterus has firmly clasped the fœtal body, such a change will often prove impossible; but in many cases, before this unfortunate chain of circumstances has occurred, the operation of podalic version with all its serious consequences to mother and child may be avoided, and a natural parturition be substituted for an unnatural one.

In 1861[1] Sims described the disease known as vaginismus, which had, however, been previously noted by Burns, Simpson, Debout, and several others, and recommended for its relief a procedure which, while it involves little risk to the patient, insures a certain removal of the disorder. This consists in ablation of the remains of the hymen and section of the tissues at the perineal extremity of the ostium vaginæ.

Several European authorities have advocated in preference to this plan forcible distention of the ostium vaginæ and modification of the local nervous hyperæsthesia by alterative applications. A comparison of the two methods at the bedside will be greatly in favour of the former.

In 1862[2] E. Noeggerath, of New York, proposed and practised the method of reduction of an inverted uterus by digital compression of both horns. He based this procedure upon the pathological fact that inversio uteri generally begins by inversion of the horns. Experiment proves the method of Noeggerath to be a valuable and reliable one, which should rank among the important contributions which have been made to this subject.

In 1867[3] Theophilus Parvin described an operation for uretero-vaginal fistula, a condition which had previously attracted little attention. This consisted in first turning the displaced distal extremity of the ureter into the bladder, and then closing the vaginal opening. The case reported was the first of this kind upon which the operator had essayed the method, and it proved entirely successful.

In 1868[4] a valuable suggestion, illustrated by a case, was

[1] Trans. Obstet. Soc. London, vol. ii. p. 356.
[2] Bulletin N. Y. Acad. Med., vol. i. p. 410.
[3] Western Journ. of Med., vol. ii. p. 603.
[4] Am. Journ. of Med. Sciences, January, 1868, p. 91.

made by T. A. Emmet for the management of cases in which partial success attends reposition of an inverted uterus. This consisted in keeping the partially replaced body within the cervix by closing the os externum uteri by silver sutures. By this method the advance gained at one sitting is not lost, and the case is better prepared than it would otherwise be for further efforts.

In 1869[1] Julius F. Miner, of Buffalo, made a valuable contribution to the management of the pedicle of tumours removed by ovariotomy. His method consisted in stripping off from the tumour the expansion of the pedicle instead of ligating and severing it. In many cases Miner's method is of inestimable value, and allows of a successful issue to cases which would otherwise prove exceedingly difficult if not impossible of management.

J. Marion Sims[2] in the same year published an important essay entitled the "Microscope in Diagnosis and Treatment of Sterility." His observations bore especially upon the deleterious effects exerted upon the vitality of the zoosperms by ichorous discharges from the endometrium. Treatment, of course, was to be directed to the eradication of the disorder which gave rise to this devitalizing secretion.

In 1870[3] Gaillard Thomas performed the operation of vaginal ovariotomy, removing an ovarian cyst the size of a large orange through an opening made through the vagina and Douglas's pouch. This was the first time that this procedure was ever advised or practised for this purpose. His patient recovered.

In 1872[4] R. Davis, of Wilkesbarre, Pa., in the same manner successfully removed an ovarian cyst weighing nine pounds. In rupturing adhesions, which were abundant, his

[1] Buffalo Med. and Surg. Journ., June, 1869, p. 418. See also American Journ. Med. Sci., Oct. 1872, p. 391.
[2] N. Y. Med. Journ., January, 1869, p. 393.
[3] Amer. Journ. Med. Sciences, April, 1870, p. 387.
[4] Trans. State Med. Soc. of Penna., 1874, p. 221.

hand was passed high up into the peritoneal cavity, the sac extending several inches above the umbilicus, and forming a tumour about the size of a pregnant uterus at seven months of utero-gestation. His patient recovered.

In 1873[1] J. T. Gilmore, of Mobile, Ala., performed the same operation successfully. The temperature of his patient never rose to 100° F.

In 1874[2] Robert Battey, of Atlanta, Ga., removed in the same way a cyst the size of a small orange. The patient rapidly recovered.

By the same method, Battey has nine times extirpated the ovaries in pursuance of a plan which will now be mentioned, and Marion Sims has done so three times.

In 1872 Robert Battey[3] published an essay advocating extirpation of the ovaries with the intent of prematurely inducing the menopause in cases in which menstruation is productive of very bad results. To use his own words, it is "an operation for the removal of the normal human ovaries, with a view to establish at once the 'change of life,' for the effectual remedy of certain otherwise incurable maladies."

Too short a time has thus far elapsed for this bold innovation to have received its just estimate. It is not saying too much, however, even now to declare that its future will probably be one of a great deal of usefulness when it has been circumscribed by proper limits and the class of cases to which it is appropriate has been clearly defined. Thus far Battey's operation has been practised in the United States

by Robert Battey 10 times. 8 recoveries. 2 deaths.
" Marion Sims 5 ". 4 " 1 "
" Gaillard Thomas 1 " 1 " 0 "

[1] N. O. Med. and Surg. Journ., Nov. 1873, p. 341.
[2] Personal communication.
[3] Atlanta Med. and Surg. Journ., Sept. 1872, p. 321

Battey[1] thus expresses himself concerning some of the important points connected with this subject:—

"I have operated in widely different circumstances. In one case the patient had amenorrhœa, convulsions, recurrent hematocele, repeated pelvic abscesses, incipient tuberculosis from pulmonary congestions, etc. Several of the cases passed under the head of ovarian neuralgia; several had intractable dysmenorrhœa with pelvic deposits of old lymph; one had ovarian insanity, etc. All had exhausted the available resources of the art to no useful purpose. *I operate upon no case that any other respectable medical man proposes to cure.* In most of my cases the full results of the menopause have not yet been developed. This is the work of many months, and sometimes two or three years are necessary to its full and perfect realization. In no case has the patient failed to realize such a degree of relief and benefit following the operation as to amply compensate her for all the pains and dangers incident thereto, to say nothing of the promise of full and ample recovery at the completion of the physiological 'change.' In two of my cases this *change* has seemed to occur at once in all its completeness; but it is always my expectation that it will occur gradually, and extending through two or even three years to its final completion. In my first case (now three years ago) the restoration to health is eminently satisfactory. It is true that she is not absolutely and perfectly well, but she is fully relieved of the convulsions, the violent periodical congestions, the hematoceles, the pelvic abscesses, etc., for which I operated. I submit to you the question in all sincerity, if I confine myself to cases where life is endangered, or where health and happiness are destroyed—cases which are utterly hopeless of other remedy this side the grave—ought the profession to demand at my hands the restoration of these forlorn invalids to a state of complete and absolute health in every particular?"

In 1873[2] John Ball, of Brooklyn, published the results of a plan of treating constrictions and tortuosities of the canal of the cervix uteri resulting from versions and flexions, by

[1] Amer. Practitioner, Oct. 1875, p. 207.
[2] N. Y. Med. Journ., Oct. 1873, p. 363.

rapid dilatation, by expanding instruments of steel. Ellinger, of Germany, has likewise adopted this heroic method, but Ball declares that he has employed it for several years, and without the knowledge that any one else was testing it. The procedure is thus described by its originator :—

"My method of procedure is first to evacuate the bowels pretty thoroughly beforehand, so as to prevent all effort in that direction for two or three days; I then place the patient upon her back, with her hips near the edge of the bed, and, when she is profoundly under the influence of an anæsthetic, I commence by introducing a three-bladed, self-retaining speculum, which brings in view the os uteri, which I seize with a double-hooked tenaculum, and draw down toward the vulva, when I first introduce a metal bougie as large as the canal will admit, followed in rapid succession by others of larger size until I reach No. 7, which represents the size of my dilator. I then introduce the dilator, and stretch the cervix in every direction, until it is enlarged sufficiently to admit a No. 16 bougie, which is all that is generally necessary. Then I introduce a hollow, gum-elastic uterine pessary, of about that size, and retain it in position by a stem, secured outside of the vulva, for about a week, in which time it has done its work, and is ready to be removed.

"During this time I keep the patient perfectly quiet, and usually upon her back, which is generally found to be the most comfortable position."

To the uninitiated this procedure appears fraught with great danger, but the originator declares that out of between twenty and thirty cases he has met with but one fatal issue. He says:—

"According to my own experience, it causes much less constitutional disturbance than the use of tents; and I think it safer even than the metrotome, and free from some serious objections to the use of the latter; as, for instance, when incisions are made through the tissues of the cervix, unless carried deep enough to prevent reunion, they must of necessity form a cicatrix, which will interfere, more or less, with the dilatation of the parts. And, when the operation does not succeed, the patient is left in a worse condition than before; while, in the rapid and forcible dilatation of the cervix, there is no sacrifice of the integrity of the parts, and, being done under the influence of an anæsthetic, there is no shock of the nervous system, and generally but little subsequent suffering."

In 1874 an important contribution to the pathology and treatment of diseases of the cervix uteri emanated from T. A. Emmet.[1] It had long been known, that, as the head of the child passed the os externum uteri, lacerations of its muscular walls often occurred; but up to this time it had not been recognized how uniformly this condition is confounded with the so-called ulceration of the cervix, and how commonly the eversion of the lips of the cervix resulting from it is mistaken for hypertrophy of the cervical tissues. Emmet advocated for this condition vivification of the edges of the lacerated parts, and approximation of them by suture. This procedure is one of most beneficent character, and one which must take rank as an important advance in gynæcological surgery.

The medical literature of the first quarter of the present century contains several allusions to an operation styled gastro-elytrotomy, a procedure intended to avoid cutting through the uterus and peritoneum, and yet allowing of the removal of the child through the abdominal walls and above the true pelvis.

This operation has attracted the attention of four obstetricians: Jorg in 1806, Ritgen in 1820, Physick in 1822, and A. Baudelocque in 1823. Kilian, in speaking of Jorg's conception of the operation, says that he merely suggested it; and even if he had performed it, his results would not have been admitted in a fair appreciation of the operation, since he did not propose avoiding the peritoneum, a prominent feature of the method. The same writer alludes to

---

[1] N. Y. Med. Journ., July, 1874.

one operation by Ritgen which ended fatally. In 1870, Gaillard Thomas, without a knowledge of the fact that he had been anticipated in the procedure, delivered in this way a living child. The operation was at that time thus described by him:—[1]

"The patient being placed upon a table, anæsthesia was produced, so as to quiet her restlessness and jactitation, with a few inhalations of ether. I then passed my hand up the vagina and dilated the cervix slowly and cautiously, so that at a three-quarter distention no injury was done to its tissue. With a bistoury I then cut through the abdominal muscles, the incision being carried from the spine of the pubis to the anterior superior spinous process of the ileum. The lips of the wound were now separated, and by two fingers the peritoneum was lifted with great readiness, so that the vagino-uterine junction was reached. The vagina was now lifted by a steel sound passed within it, and cut, and the opening thus made was enlarged by the fingers. The cervix was then lifted into the right iliac fossa by the blunt hook, while the fundus was depressed in an opposite direction. I then passed my right hand into the iliac fossa, and introduced two fingers into the uterus, while the left hand, placed on the outer surface of the uterus, depressed the pelvic extremity of the fœtal ovoid. The knee was readily seized, and delivery easily and rapidly accomplished."

In 1876,[2] Alexander J. C. Skene, of Brooklyn, performed this operation with a brilliancy of result never before attained by any one. The patient was a small rachitic woman, aged thirty-one years, who had been three times delivered, once by craniotomy and twice by premature delivery at the seventh and eighth months. One of the last two children had lived a few minutes, and one for several months. In her fourth pregnancy Dr. Skene let gestation advance to full term; then, finding an arm and the cord

[1] Amer. Journ. Obstet., vol. iii. p. 125.
[2] Ibid., vol. viii. p. 636.

presenting, he performed gastro-elytrotomy, saving the mother and a vigorous child weighing ten pounds. Both made a perfect recovery.

This completes the list of those contributions to obstetrics and gynæcology on the part of this country which appear to be especially marked by originality and by practical utility. But how difficult is it to decide what really deserves the credit of original conception? "Is there anything whereof it may be said—See, this is new? it hath been already of old time which was before us. There is no remembrance of former things." As the husbandman turns up to the light and brings into activity and usefulness the mould which, though buried for ages, was in by-gone times ploughed by his predecessors, so do the seekers after new ideas bring to light the thoughts of those whose discoveries have been long ago forgotten. Who is to decide how long a time must intervene between the periods of successive discovery to warrant for the latest aspirant the claim of originality?

The peculiar features of the contributions just enumerated seem to warrant their arrangement in a special category, but this does not argue in them greater value than that attaching to those of somewhat different character which come to be considered now. Indeed, some of the latter type have exerted a more powerful and widespread influence than many of the former, and have been productive of greater good to medicine and humanity.

In June, 1842, Jos. Warrington's "Obstetric Catechism" which for a time was used as an epitome of the subject of midwifery by students, appeared.

The year 1843 was marked by the appearance of an essay[1] which was productive of a great deal of good, from

---

[1] New England Quarterly Journ. of Med. and Surg., April, 1843, p. 503.

the pen of the eminent poet-physician, Oliver Wendell Holmes. At that period the then authoritative works upon obstetrics, those of Dewees and Meigs, both maintained the non-contagiousness of puerperal fever. Holmes boldly joined issue upon this momentous point, and, although devoting much less attention to this department than the authors mentioned, his observations led him to a more correct conclusion.

In 1845 an important contribution to a subject which even now has received little attention, was made by Isaac E. Taylor, of New York, in an essay upon Rheumatism of Uterus and Ovaries.[1] In this some striking cases of rheumatic disorder of the muscular structure of the pregnant uterus were recorded.

During this year W. L. Atlee published a synopsis[2] of 101 ovariotomies, and an essay upon Intra-Uterine Fibroids.[3]

During the next year an essay appeared from Samuel Kneeland, Jr.,[4] of Boston, maintaining a close relationship between epidemic erysipelas and puerperal fever. It is well known how much favour this view has since met with.

The year 1846 was marked by a discovery in this country which may be said to overshadow any other of its contributions to medicine. I allude to anæsthesia, discovered by Horace Wells, a dentist of Connecticut, and subsequently made practicable and useful by W. T. G. Morton, likewise a dentist, of Boston. Only the relations of this subject to obstetrics and gynæcology find legitimate place in this essay.

In January, 1847, anæsthesia by ether was first employed for assuaging the pains of labour by Prof. Simpson, of

---

[1] Amer. Journ. Med. Sci., July, 1845, p. 45.
[2] Ibid., April, 1845, p. 309.
[3] Trans. Amer. Med. Assoc., vol. iii. p. 380.
[4] Amer. Journ. Med. Sci., April, 1846, p. 324.

Edinburgh; in April of the same year it was employed in this country, by Dr. N. C. Keep, of Boston; and in May, by Dr. Channing, of Boston, in a case of instrumental labour. The introduction of this beneficent agent into the lying-in chamber constitutes an era in the history of obstetrics. It is somewhat singular that after the discovery of anæsthesia in this country; after the prediction, long before its discovery, by one of America's greatest physicians, that[1] "a medicine would be discovered that should suspend sensibility altogether, and leave irritability, or the powers of motion, unimpaired, and thereby destroy labour-pains altogether;" after it had been employed here in hundreds of cases for surgical operations, this link of the chain should have been forged by a European. Yet such was the case, and far be it from any American to begrudge him one atom of the glory which he deserves, or to endeavour to dim its lustre by "faint praise."

Boston was the field in which the first demonstrations of anæsthesia, as an agent of practical value, were made, and there appeared the first and most ardent advocate of its use in obstetrics in this country. Dr. Walter Channing was elected to the chair of obstetrics in Harvard, in 1833, and was recognized as a leader in this department, both from his teachings and writings. He warmly espoused the subject, and in 1848 published a treatise on Etherization in Childbirth, illustrated by 581 cases. This volume numbers 400 pages, and served to bring the subject fully before the whole civilized world. So well did it serve its purpose that no similar work has since appeared either from an American or European author.

What a striking contrast is presented between the rapid acceptance of this discovery by the whole medical world and the tardy, unwilling, bitter reception of ovariotomy! The

---

[1] Rush, Med. Inquiries and Observations, 3d ed., vol. iv. p. 376.

first patient in Boston submitted to operation under anæsthesia, was etherized by Morton in October, 1846. Writing in April, 1847, Hayward declares that ether "has probably been used in this way by several thousand individuals in this city within the last six months," and, in 1848, Channing,[1] of the same city, illustrates the utility of this agent in the lying-in chamber alone by the citation of over five hundred cases.

In 1847, I. E. Taylor contributed an essay drawing attention to the causation of exophthalmos and enlargement of the thyroid gland by excessive lactation; and Fordyce Barker one upon diseased states of the uterine neck.

In 1850 the first attempt at the establishment of an obstetric clinic in this country was made by J. P. White, of Buffalo. In furtherance of this mode of instruction, the act of human parturition was displayed ocularly to some sixteen students, the professor explaining its features during its accomplishment. A perfect storm of popular, and to a certain extent of professional indignation, was excited by this, which was only stemmed by the dignified and bold attitude of the united faculty of the University of Buffalo, and the support lent by enlightened obstetricians though-out the land.

The whole subject was fully brought out in the trial of the People v. Dr. H. N. Loomis, a report[2] of which to-day constitutes a curious episode in the medical literature of the century. In this will be found a letter signed by seventeen physicians, characterizing the demonstration as "wholly unnecessary, and grossly offensive, alike to morality and common decency."

---

[1] Etherization in Childbed, Boston, 1848.
[2] Jewett, Thomas & Co., Buffalo, 1850.

During the next year a full synopsis[1] of all the known cases of ovariotomy which had up to that time been performed appeared from W. L. Atlee. This embodied 222 cases, and constituted the most valuable statistical table which had yet appeared.

In the same year a masterly essay upon the Corpus Luteum of Menstruation and Pregnancy[2] was submitted to the American Medical Association, and was awarded the prize. Its author was John C. Dalton, of New York.

In 1853 Thomas F. Cock published a Manual of Obstetrics,[3] a *multum in parvo* of the most reliable maxims in that art, which even now constitutes the *vade mecum* of many of our students.

Two years afterwards a paper[4] was read by Fordyce Barker before the New York Academy of Medicine upon the Treatment of Puerperal Convulsions, which fully presented all that was then known upon a subject which has since called forth so much discussion. In the same year R. A. F. Penrose published an interesting and valuable essay upon a Case of Triplets, with the Mechanism of Labour.[5]

In the same year James Deane, of Massachusetts, published an essay upon "The Hygienic Condition of the Survivors of Ovariotomy," which was particularly valuable at a time when this operation was being weighed in the balance and its beneficent results doubted by many of the most sincere investigators.

In 1856[6] there appeared the most exhaustive and valuable essay upon ovariotomy which had yet been published. This was the prize essay of Geo. H. Lyman, of Boston,

---

[1] Trans. Amer. Med. Assoc., vol. iv. p. 286.
[2] Ibid., p. 547.   [3] W. Wood & Co., N. Y.
[4] Transactions, vol. i. p. 273.   [5] Med. Exam., Feb. 1855, p. 77.
[6] Publications of Mass. Med. Soc., May, 1865.

entitled, "History and Statistics of Ovariotomy, and the Circumstances under which this Operation may be regarded as Safe and Expedient." It appeared at a most opportune moment, and, characterized as it was by a fair and manly spirit, a remarkable degree of accuracy, and entire absence of narrow and prejudiced views, it did a great deal of good in reference to the important subject with which it dealt. Although twenty years have elapsed since its publication, it can still be read with profit and be regarded as a safe guide in reference to many essential points.

During the years 1848 and 1856 there appeared in the *American Journal of the Medical Sciences*[1] some very valuable essays of statistical character upon rupture of the uterus by J. D. Trask, of Astoria, Long Island. These were valuable by the faithfulness and accuracy which characterized them, and the thoroughness with which the subject was treated. The same author has now nearly ready for publication a paper bringing the subject down to the present day.

In the latter of these years I. E. Taylor, in a report of Two Cases of Recto-Vaginal Fistula, cured by a New Operation,[2] advocated severance of the sphincter ani in such cases after the manner of Rhea Barton. During the succeeding year two valuable papers appeared, one by Emil Noeggerath upon Metastatic After-Pains,[3] and another by J. Marion Sims, upon Silver Sutures in Surgery.[4]

In 1858 J. Foster Jenkins,[5] of Yonkers, made an important contribution to the literature of the subject of spontaneous umbilical hemorrhage in the newly born, and William

---

[1] N. S., vol. xv. pp. 104 and 383, and xxxii. p. 81.
[2] N. Y. Med. Journ. [3] N. Y. Journ. Med., May, 1857, p. 287.
[4] Trans. N. Y. Acad. Med., Nov. 1857.
[5] Trans. Amer. Med. Assoc., vol. xi. p. 263.

Read, of Boston, one upon the influence of the Placenta upon the Development of the Uterus during Pregnancy.[1]

During the following year three essays well worthy of note appeared; two by Noeggerath, upon the Local Disinfecting Treatment of the Cavity of the Uterus for the Treatment of Puerperal Fever,[2] and on the Operation of Turning by External Manipulations;[3] one by Sims upon Amputation of the Cervix, Stump covered with Vaginal Membrane.[4]

In 1861 William Read published a paper[5] upon The Formation of Knots in the Umbilical Cord, and Fordyce Barker[6] one on the Use of Anæsthetics in Midwifery. The latter of the subjects was one requiring at that time all the light which could be shed upon it by conscientious observers.

In this year there appeared[7] an interesting paper by Samuel R. Percy, of N. Y., demonstrating the tenacity of vitality possessed by the human zoosperm. His statements are here given in his own words:—

"I was called to attend a lady with uterine disease, but I considered it best to postpone all treatment, as on the next week her husband would leave home to be absent two or three months. On the Monday following he left, but she did not call upon me until a week from the day following. On examination with the speculum I found a mass of what I supposed to be muco-purulent matter, proceeding from the os uteri. Wishing to ascertain its character, I examined it with the microscope, and was surprised to find that it was semen, and that it con-

---

[1] Am. Journ. Med. Science, April, 1858, p. 309.
[2] Contrib. to Midwifery and Dis. of Women, New York, 1859.
[3] N. Y. Journ. of Med., Nov. 1859, p. 329.
[4] Trans. N. Y. State Med. Society, 1861, p. 367.
[5] Am. Journ. Med. Sci., Oct. 1861, p. 381.
[6] Trans. N. Y. Acad. Med., vol. ii. p. 251.
[7] Amer. Med. Times, March, 1861, p. 160.

tained living spermatozoa and many dead ones. Communicating in a proper way my discovery, I questioned her as to the time of her last intercourse with her husband. It was on the Monday morning before leaving, nearly eight and a half days previous. I would stake my reputation on her honour." Dr. Percy further says: "Knowing that the zoosperms of the frog are frequently found living days after the frog's death, and even when it has been frozen, I can conceive no reason why human spermatozoa may not retain their vitality for some time, especially when protected by warmth and placed in the situation where nature designed them. But to test this matter, I placed some semen in the lower part of a piece of moistened membrane, tied it, and placed it within the vagina of a mongrel bitch. Upon removing it, upon the sixth day, most of the zoosperms were possessed of vitality, though there were many dead ones. These facts may have an important bearing in a medico-legal way."

During the next year the subject of Pelvic Hæmatocele began to attract considerable attention in America. From the year 1831, in which it was first described by Récamier, cf Paris, it had not ceased to attract considerable attention in France, and between that period and 1858 Bernutz, Vignes, Nelaton, Gallard, and Voisin had written their well-known essays upon it. Up to this year, however, only one case had been reported amongst us, and it constituted an era in the subject for three essays to appear in one year. One was by John Byrne, of Brooklyn; one by Fordyce Barker; and one by E. Noeggerath. All these were read before the N. Y. Academy of Medicine, and appeared in its Transactions.

During this year I. E. Taylor published a valuable essay[1] upon the non-shortening of the supra and infra-vaginal portion of the cervix uteri up to the full term of gestation. In this the author contested the views of Stoltz, of Stras-

[1] Am. Med. Times, vol. iv. p. 342.

bourg, to the effect that gradual expansion of the cervical canal during the latter months of pregnancy effaced or obliterated that portion of the uterus.

In 1863, H. R. Storer,[1] of Boston, added to the literature of the subject of anæsthesia in midwifery and medical surgery an essay of considerable value; Barker one upon Albuminuria[2] as affecting Pregnancy, Parturition, and the Puerperal State; and E. N. Chapman a report[3] entitled a Selection of Remarkable Cases.

The next year saw the publication of two able papers upon Ovarian Tumours and Ovariotomy,[4] by E. R. Peaslee; an essay upon Spinal Irritation,[5] by Charles F. Taylor; and an excellent treatise upon Chronic Inflammation and Displacement of the Unimpregnated Uterus, by W. H. Byford, of Chicago.

In 1865, T. A. Emmet published upon the Treatment of Dysmenorrhœa and Sterility,[6] resulting from Anteflexion of the Uterus, and upon the Radical Operation for Procedentia;[7] I. E. Taylor upon Placenta Prævia;[8] and Peaslee[9] gave Statistics of 150 Cases of Ovariotomy.

The work of Byford, mentioned as having appeared in 1864, had already met with so brilliant a success that it now reappeared, enlarged and improved, under the title of the Medical and Surgical Treatment of Women.

During the next year, I. E. Taylor[10] reported sixty cases

[1] Boston Med. and Surg. Journ., vol. lxix. p. 249.
[2] Bulletin N. Y. Acad. Med., vol. ii. pp. 36 and 67.
[3] Med. and Surg. Reporter, Phila.
[4] Bull. N. Y. Acad. Med., vol. ii. p. 226.
[5] Trans. N. Y. State Med. Soc., 1864, p. 126.
[6] New York Med. Journ., June, 1865, p. 205.
[7] Ibid., April, 1865, p. 1.
[8] Trans. N. Y. State Med. Soc., 1865, p. 62.
[9] Am. Journ. Med. Sci., Jan. 1865, p. 89.
[10] Trans. N. Y. State Med. Soc., 1866, p. 97.

of recto-vaginal and recto-labial fistulæ treated by the plan already mentioned, and Emmet[1] published an essay upon Atresia Vaginæ.

This year was specially marked by the appearance of a work which more profoundly aroused the gynæcologists of Europe, as well as of America, than any other which had appeared since those of Bennet and Simpson. This was a work entitled Clinical Notes on Uterine Surgery, by J. Marion Sims. The clear, forcible, and persuasive style of this work, the record of successful operations which it contained, and the stamp of earnest and original thought which it bore upon every page, served to give it a circulation which demanded its translation into almost all the modern languages of Europe, and to make it an essential in the library of every progressive gynæcologist. Ten years have elapsed since its publication, and yet it may safely be stated that no work now extant constitutes a more perfect guide to the gynæcological surgeon.

In this year appeared, too, an excellent treatise of over fifty pages upon Vesico-vaginal Fistula, by M. Schuppert, of New Orleans. This contained an exhaustive *resumé* of the history of the operation, was fully illustrated, and embodied the extensive experience of one who has made himself well known as a successful operator in this field of surgery.

In this year, also, especial attention was called to the subject of extirpation of the uterus for fibroids, by the publication of a successful case, by H. R. Storer,[2] in which this organ and both ovaries were removed. This grave procedure, recommended, but never practised, as early as 1787, by Wrisberg, was in the present century performed by Clay, of Manchester, and Kœberlé, of Strasbourg. In

---

[1] Richmond Med. Journ., Aug. 1866, p. 89.
[2] Am. Journ. Med. Sci., Jan. 1866, p. 110.

1854, the first operation was performed in this country for this purpose, by Kimball,[1] of Lowell, the tumour weighing six pounds, and the patient recovering. He has been followed by Burnham, Cutter, Peaslee, Darby, Sims, Atlee, Wood, Sands, Buckingham, Storer, Hackenberg, Weber, Thomas, Chadwick, and others. The statistics of the procedure in this country have not been collected, but it is safe to say that no such results can be reported as have recently come to us from Paris, where M. Péan has met with a success of seven out of nine, or an equivalent of seventy-eight out of one hundred. Kimball has thus far performed ten operations, with four recoveries and six deaths. In New York city the operation has been repeatedly performed, but never yet with a favourable issue.

In 1867, Dr. E. D. Miller,[2] of Boston, published an essay introducing into practice the scarification of the lining membrane of the body of the uterus, and described an instrument for performing this operation; and a valuable paper was read before the American Medical Association by Stephen Rogers,[3] of New York, advocating gastrotomy after rupture of the cyst of extra-uterine pregnancy, for the purpose of ligating bleeding vessels, and thus giving the patient a chance for life. As early as 1849 this course had been suggested by W. W. Harbert in the *Western Journal of Medicine and Surgery;* but to Rogers belongs the credit of pressing the claims of the idea upon the profession in a way to attract to it the grave attention which it deserves.

Montrose A. Pallen, formerly of St. Louis, now of New York, read in the same year an interesting paper on the Treatment of Certain Uterine Abnormities, before the American Medical Association, and published a *Resumé*[4] of forty-

---

[1] Bost. Med. and Surg. Journ., May, 1855, p. 249.
[2] Ibid., March, p. 133.
[3] Trans. Am. Med. Assoc. 1867, vol. xviii. p. 85.
[4] Humboldt, Med. Archives, 1867, vol. i. p. 7.

six operations for dysmenorrhœa by the division of the cervix uteri.

Wm. T. Lusk[1] also made a contribution entitled Uræmia, a Common Cause of Death in Uterine Cancer.

In this year H. Lenox Hodge,[2] in a case of tubo-uterine pregnancy, performed a very remarkable and successful operation for removal of the fœtus. The pregnancy had advanced to the eighth month, and a thin septum divided the true and unoccupied uterus from the adjoining vicarious one so as to prevent delivery. Hodge cut through this, and delivered the child *per vias naturales*. The child lived about ten hours, and the mother recovered.

In 1868 the first journal ever devoted especially to the interest of obstetrics and gynæcology in America appeared in New York. The establishment and early maintenance of this excellent quarterly, styled *The American Journal of Obstetrics and Diseases of Women and Children*, were entirely the results of the energy and enterprise of a single member of the profession, B. F. Dawson. After eight years of existence it has established its right to be considered one of the most valuable periodicals of the country, and under its present editor, Paul F. Munde, fully maintains its position.

During this year there appeared three works in this department of medicine; first, the Obstetric Clinic of George T. Elliot, classic in style, and replete with the wise counsels of a master in the obstetric art; second, a Treatise upon Vesico-vaginal and Recto-vaginal Fistulæ, by T. A. Emmet; and third, a Practical Treatise upon the Diseases of Women, by Gaillard Thomas.

Two good papers likewise appeared, one upon Intra-

---

[1] N. Y. Med. Journ., June, 1867, p. 205.
[2] Parry, op. cit. p. 266.

uterine Injections, by M. A. Pallen,[1] and one upon the Treatment of the Uræmic Convulsions of Pregnancy by Morphia, by F. D. Lente.[2]

In 1868, H. R. Storer,[3] of Boston, advocated inclosing the pedicles of ovarian tumours in the abdominal walls. This method, which he styled "pocketing the pedicle," consisted of fixing it in the abdominal opening and completely covering it by the cutaneous tissues.

The year 1869 was rich in essays of considerable value. Chief among these may be mentioned one upon Ovariocentesis Vaginalis,[4] and another upon Chronic Metritis in its relation to Malignant Disease of the Uterus, by Noeggerath; one by Wm. Goodell,[5] of Philadelphia, upon Concealed Accidental Hemorrhage of the Gravid Womb; one upon the Surgery of the Cervix in connection with the treatment of certain Uterine Diseases, by T. A. Emmet;[6] one upon Hypodermic use of Ergot in Post-partum Hemorrhage, by F. D. Lente;[7] one upon Face Presentations, by I. E. Taylor;[8] one upon Intra-uterine Injections, by Joseph Kammerer;[9] one by J. G. Pinkham,[10] of Lynn, on Scarification of the Fundus Uteri in Chronic Metritis and Endometritis, which had been previously advised by Miller; one by H. R. Storer[11] upon a Method of Exploring and Operating upon the Female Rectum by Eversion of the Anterior Rectal Wall by a finger in

---

[1] St. Louis Med. and Surg. Journ., July, 1868, p. 294.
[2] Med. Record, April 15.
[3] Ibid., Jan. 15, 1868, p. 519.
[4] Amer. Journ. Obstetrics, May and November.
[5] Ibid., vol. ii. pp. 1, 505, and 610.
[6] Ibid., February, 1869, p. 339.
[7] N. Y. Med. Record, vol. iv. p. 411.
[8] N. Y. Med. Journ., November, 1869, p. 125.
[9] Read before Co. Med. Soc.
[10] Journal Gynæcological Society, Boston, vol. i. p. 23.
[11] Ibid., vol. i. p. 24.

the Vagina; and one upon the Pathological Sympathies of the Uterus, by V. A. Taliaffero,[1] of Ga.

The literature of fibro-cystic tumour of the uterus is very recent. In 1869 Koeberlé, of Strasbourg, tells us that only fourteen cases had been recorded, and of these two were discovered *post-mortem*. In that year C. C. Lee,[2] of New York, collected nineteen cases, and published them in an interesting paper.

In the same year[3] Gaillard Thomas published the account of a case of inversion successfully reduced by dilatation of the constricting neck through an opening in the abdomen made by section through its walls. This procedure has not met with favour, and has not since that time been repeated by any one but its author.

During this year a society, which exerted considerable influence in arousing attention to the subject of Gynæcology in New England, was formed in Boston, chiefly through the exertions of Horatio R. Storer, and called the Gynæcological Society of Boston. Before the year had expired a journal emanated from this society styled the Journal of the Gynæcological Society of Boston. It now no longer exists, but during its period of publication it exercised a decided influence in this department of medicine.

Societies devoted to obstetrics and gynæcology have likewise been established in Louisville, Philadelphia, and New York. They are still in active operation, and furnish in their proceedings and reports a valuable fund of information to the general reader of the medical periodicals of the country.

We now arrive at the commencement of the present decade, and during the six years of it which have now expired,

---

[1] Journ. Gynæc. Soc. Bost., vol. i. p. 341.
[2] Med. Record, vol. iv. p. 495.
[3] Amer. Journ. Obstetrics, November, 1869, p. 423.

so numerous have been the contributions to this department, that only a small proportion, consisting of the most valuable, can be noticed. During the first year of this period, there appeared Byford's Treatise on the Theory and Practice of Obstetrics, the first systematic work upon this subject which had appeared since that of Bedford, which is elsewhere noticed. An excellent paper likewise appeared from C. C. P. Clark,[1] of Oswego, upon the Management of the Obstetric Forceps, replete with the sagacious observations of an original and candid observer; and an important essay upon Anal Fissure in Women, by H. R. Storer.[2]

The next year produced a carefully prepared and interesting essay by Wm. Goodell,[3] entitled A Critical Inquiry into the Management of the Perineum during Labour; one of great practical value by the late John S. Parry[4] upon Sudden Enlargement of Ovarian Cysts from Hemorrhage into them; one upon Dysmenorrhœa and its Treatment by M. A. Pallen;[5] a report of a Case of Simultaneous Intra- and Extra-uterine Pregnancy going to Full Term, by S. Pollak;[6] a paper upon Mechanical Treatment of Displacement of Unimpregnated Uterus, by George Pepper, of Philadelphia;[7] a very valuable essay upon Placental Extraction and Placental Expression, by Parvin;[8] and an equally valuable one by Nathan Bozeman[9] upon Urethrocele, Catarrh, and Ulceration of the Bladder in Females.

The next year was marked by the appearance of three

[1] Trans. N. Y. State Med. Soc., 1870, p. 249.
[2] Journal Gynæcological Society, Boston, vol. ii. p. 221.
[3] Amer. Journ. Med. Sci., Jan. 1871, p. 53.
[4] Amer. Journ. Obstetrics, Nov. 1871, p. 454.
[5] Missouri Med. and Surg. Report.
[6] St. Louis Med. and Surg. Journal, May, 1871, p. 193.
[7] Amer. Journ. Obstetrics, Aug. 1871, p. 258.
[8] Trans. Ind. State Med. Soc., 1871, p. 11.
[9] Amer. Journal Obstetrics, Feb. 1871, p. 636.

works devoted to this department of medicine; first, one upon Ovarian Tumours, their Pathology, Diagnosis, and Treatment, by E. R. Peaslee; second, one upon Hysterology, by E. N. Chapman; and third, one upon Electro-cautery in Uterine Surgery, by John Byrne. The first of these is certainly the most systematic and complete treatise which has thus far appeared upon this subject, and the last, although small in dimensions, deals exhaustively with the important matter upon which it touches.

During this year there appeared a remarkable essay upon Latent Gonorrhœa in Females, by Noeggerath. In this the author strongly assumes the position which is here announced in his own words.[1]

"I have undertaken to show that the wife of every husband who, at any time of his life before marriage, has contracted a gonorrhœa, with very few exceptions, is affected with latent gonorrhœa, which sooner or later brings its existence into view through some one of the forms of disease about to be described. .... I believe I do not go too far when I assert that, of every 100 wives who marry husbands who have previously had gonorrhœa, scarcely 10 remain healthy; the rest suffer from it or some other of the diseases which it is the task of this paper to describe. And, of the ten that are spared, we can positively affirm that in some of them, through some accidental cause, the hidden mischief will sooner or later develop itself."

The disorders supposed by the author to result from latent gonorrhœa are perimetric inflammations, both acute and chronic; ovaritis; and catarrh of the genital tract.

In the same year Parry[2] published an essay upon the comparative merits of craniotomy and Cæsarean section in pelves with a conjugate diameter of $2\frac{1}{4}$ inches or less.

The idea of draining the peritoneal cavity by creating an opening *per vaginam* into its most dependent portion, the

---

[1] Published in Bonn.
[2] Amer. Journ. Obstet., Feb. 1873, p. 644.

pouch of Douglas, has often presented itself to the minds of ovariotomists. As early as 1855[1] the practice was adopted by Peaslee, and subsequently by Kimball, of Lowell; W. W. Green and Tewkesbury, of Portland; Miner, of Buffalo; Thomas, of New York, and others. This plan of accomplishing drainage of the peritoneal sac has by no means met with general approval or adoption. Nor is it probable that it will ever do so, for between the perito. neum and vagina there is an interspace which is filled by areolar and adipose tissue into which an escape of putrid fluid must often enter and from which it would readily be absorbed.

In 1872 Marion Sims[2] revived the method and by passing into the peritoneum, through the vagina, tubes of small calibre admitting of perfect drainage, and the use of disinfectant injections he hoped to overcome more perfectly than had hitherto been done, the fatal consequences of septicæmia. The reviver of this plan of drainage still has sanguine hopes of its success, and commonly resorts to it.

In the year 1873, Thomas M. Drysdale of Philadelphia, after a careful and conscientious study of the subject, described a peculiar characteristic corpuscle as contained in ovarian fluid. This he regarded as diagnostic of ovarian cystoma. He sums up his views upon the matter in these words :—

"I claim then, that a granular cell has been discovered by me in ovarian fluid, which differs in its behaviour with acetic acid and ether from any other known granular cell found in the abdominal cavity, and which, by means of these reagents, can be readily recognized as the cell which has been described; and further, that by the use of the microscope, assisted by these tests, we may distinguish the fluid removed from ovarian cysts from all other abdominal dropsical fluids."

[1] Handyside, of Edinburgh, first did this in 1846.
[2] New York Med. Journ., Dec. 1872, p. 561.

These views are by no means generally accepted by microscopists, but their author feels sure of his position, and W. L. Atlee, many of whose diagnoses have in great degree rested upon them, has full faith in its correctness.

In this year a remarkable paper entitled, "How do the Spermatozoa enter the Uterus,"[1] by Joseph R. Beck, of Indiana, appeared. The author, meeting with a female patient, the subject of prolapsus uteri, who was so excitable as to have the sexual orgasm produced by digital examination, examined visually as this occurred, and thus reports what he saw: "The os and cervix uteri had been firm, hard, and generally in a normal condition, with the os closed so as not to admit the uterine probe without difficulty; but immediately the os opened to the extent of fully an inch, made five or six successive gasps, drawing the external os into the cervix each time powerfully, and at the same time becoming quite soft to the touch. All these phenomena occurred within the space of twelve seconds' time certainly, and in an instant all was as before; the os had closed, the cervix hardened, and the relation of the parts had become as before the orgasm." Similar observations had been previously made by Sitzmann,[2] of Germany, and published in 1846.

In the next year W. L. Atlee gave to the profession a work entitled General and Differential Diagnosis of Ovarian Tumours, with special reference to the operation of ovariotomy, and occasional pathological and therapeutical considerations. This embodied his vast experience, and recorded the results of his numerous operations for the removal of tumours of the uterus and ovaries.

In this year likewise appeared the work of D. Hayes Agnew, of Philadelphia, upon Laceration of the Female

---

[1] St. Louis Med. and Surg. Journ., New Series, vol. ix. p. 449.
[2] A. Flint, Physiology of Man, vol. v. p. 339.

Perineum and Vesico-vaginal Fistula, their history and treatment. This likewise was the production of a man of mature thought and great experience and knowledge of the subject with which he dealt. To this author the profession is indebted for a great deal of honest and valuable labour in reference to the surgery of the female genital organs.

A lengthy report, in book form, of the Columbia Hospital for Women, in Washington, D. C., was made by J. H. Thompson, and a valuable essay was published by T. A. Emmet,[1] upon Laceration of the Perineum, involving the Sphincter Ani, and an Operation for securing Union of the Muscle. Emmet in this essay urges upon operators the necessity of inserting the first suture low down, on a level at least with a horizontal line running along the lowest edge of the anus, so as to lift the ends of the broken muscle up, and cause them to approximate. This constitutes the pivotal point of the operation.

In this year, also, appeared an essay by W. T. Lusk,[2] on the Etiology and Indication for Treatment of Irregular Uterine Action during Labour.

The year 1874 was very prolific in contributions to this department. In it appeared a work which has met with great and deserved success by Fordyce Barker, upon the Puerperal Diseases, and an essay by the same author,[3] upon The Age when the Capacity for Child-bearing ceases.

In this year two articles appeared from one of America's greatest ovariotomists, Gilman Kimball,[4] of Lowell, upon Pelvic Drainage after Ovariotomy; a noteworthy report by the late A. K. Gardner, of a case in which ten quarts of urine were at one operation removed from the female blad-

---

[1] Med. Record, March, 1873, p. 121.
[2] N. Y. Med. Journ., June, 1873, p. 561.
[3] Phila. Med. Times, vol. v. p. 161.
[4] Boston Med. and Surg. Journ., 1874, N. S., vol. xiii. p. 517, and vol. xiv. pp. 132, 272.

der; a paper by H. Lenox Hodge,[1] upon Injection of Tincture of Iodine into the Cavity of the Uterus in Hemorrhage after Delivery; a report by Goodell[2] on The Means employed at Preston Retreat for the Prevention and Treatment of Puerperal Diseases; a most valuable and masterly essay on the Mechanism and Treatment of Breech Presentations, by R. A. F. Penrose;[3] a description of an operation styled Vagino-cerviplasty,[4] a substitute for amputation of the cervix uteri in certain forms of intra-vaginal elongation, by Pallen; an additional paper, on The Physiological Lengthening of the Cervix Uteri at, before, during, and after Delivery, by I. E. Taylor;[5] one by Marion Sims,[6] upon Enucleation of Intra-uterine Fibroids, and one upon Erysipelas in Child-bed Fever, by Thomas C. Minor, of Cincinnati.

In the same year there appeared a work, small in proportions but powerful in style and effect, from the pen of Edward H. Clarke, entitled Sex in Education. Few works in modern times upon medical topics have so thoroughly succeeded in arousing the attention of the community for whose benefit they were undertaken.

In 1875 James D. Trask,[7] of Astoria, N. Y., published an essay upon Injection of Tincture of Iodine into the Cavity of the Uterus in Hemorrhage after Delivery; J. R. Chadwick,[8] of Boston, one upon Injection of Nutritious or Cathartic Fluids into the Intestines through the abdominal

[1] Am. Journ. Obstet.
[2] Amer. Sup. to Obstet. Journ. of Great Britain, 1874, July, p. 49, and August, p. 65.
[3] Ibid., March, 1874, p. 177.
[4] Amer. Journ. Obstet., Feb. 1875, p. 604.
[5] Ibid., May, 1874, p. 119.
[6] N. Y. Med. Journ., April, 1874, p. 337.
[7] Am. Journ. Obstet., Feb. 1875, p. 613.
[8] Ibid., Nov. 1875, p. 399.

walls by means of an aspirator needle when the stomach proves entirely intolerant; A. D. Sinclair,[1] of Boston, one upon Manual Dilatation of the Os Uteri; Noeggerath[2] one upon Vesico-vaginal and Vesico-rectal Touch, a new method of examining the Uterus and Appendages; and Goodell,[3] a Clinical Memoir upon Turning in Pelves narrowed in the Conjugate Diameter, and another[4] upon The Management of Head-last Labours.

In the same year Wm. H. Byford[5] read before the American Medical Association an able report upon The Treatment of Uterine Fibroids by Ergot (Hildebrandt's method); F. D. Lente[6] and Alex. Murray[7] published essays advocating the use of electricity in arresting post-partum hemorrhage; Parry[8] one upon The Use of the Hand to correct unfavourable presentations and positions of the child during labour; and another[9] upon The History of an Outbreak of Puerperal Fever at the Philadelphia Hospital, characterized by diphtheritic deposits on wounds of the genital organs; M. B. Wright[10] one upon Obliquities of the Gravid Uterus; and Lusk,[11] a valuable report upon The Genesis of an Epidemic of Puerperal Fever. A very interesting and valuable paper appeared during this year from H. F. Campbell,[12] of Georgia, upon Position, Pneu-

[1] Boston Med. and Surg. Journ., Feb. 1875, p. 117.
[2] Am. Journ. Obstet., May, 1875, p. 123.
[3] Ibid., Aug. 1875, p. 193.
[4] Phila. Med. Times, May.
[5] Trans. Am. Med. Assoc., vol. xxv. p. 173.
[6] Am. Journ. Obstet., Nov. 1875, p. 518.
[7] Psycholog. and Med.-Legal Journ., June, 1875, p. 345.
[8] Am. Journ. Obstet., May, 1875, p. 138.
[9] Am. Journ. Med. Sci., Jan. 1875, p. 46.
[10] The Clinic, vol. ix. p. 301.
[11] Am. Journ. Obstet., Nov. 1875, p. 369.
[12] Read before Georgia Medical Association, April.

matic Pressure, and Mechanical Appliance in Uterine Displacements. The author advocates replacement of uteri affected by posterior displacement by the assumption on the part of the patient of the knee-chest position, and the introduction of an open glass tube by herself into the vagina while this position is maintained. He declares that the position, favouring as it does gravitation of the uterus and other viscera forwards, aided by the entrance of air into the vagina by the glass tube, will commonly effect reposition of the displaced organ.

During the same year a faithful Report[1] upon Obstetrics and Gynæcology was made by Wm. T. Howard, to the Medical and Chirurgical Faculty of Maryland; an essay upon Ichthyosis of the Tongue and Vulva was published by R. F. Weir;[2] one upon Menstruation and the Law of Monthly Periodicity, by J. Goodman,[3] of Louisville; and one by D. Warren Brickell, of N. O., upon Rupture of the Perineum,[4] with a description of a new operation.

S. S. Todd,[5] of Kansas City, published a good *resumé* of the subject of Anæsthetics in Labour, embodying the views of many prominent obstetricians in this country and in Europe; and Thomas[6] a case of Tubal Pregnancy treated by incision into the sac by the galvano-caustic knife, and immediate removal of fœtus and placenta through the incision thus made.

H. L. Byrd[7] published the details of a new plan of artificial respiration to be practised upon the neonatus. This

---

[1] Transactions Med. and Chir. Fac. of Med. 1875, p. 73.
[2] N. Y. Med. Journ., March, 1875, p. 240.
[3] Richmond and Louisville Med. Journ., vol. xx. p. 553.
[4] Amer. Journ. of Med. Sciences, April, 1875, p. 322.
[5] Trans. Med. Association of Missouri, 1875, p. 37.
[6] N. Y. Med. Journ., June, 1875, p. 561.
[7] Med. Record, July 31, 1875, p. 519. His first article on this subject appeared in 1870.

consists in the artificial production of the inspiratory and expiratory efforts by alternately bending the trunk of the child, held in the two palms, very much backwards and forwards. As the head and shoulders fall below a horizontal line passing through the operator's hands placed under the infant's loins, and the legs and pelvis below the same line on the other side, air rushes into the lungs by reason of the recession of the diaphragm and separation of the ribs. Then as the diaphragm is pushed upwards, and the ribs approximated by the anterior bending of the trunk, so that the child's knees approach the chin, the air is expelled.

An essay appeared during this year also from Ellwood Wilson[1] upon Version in contracted Pelvis. Controversial in style, it demonstrated the truth of the aphorism "ex collisione, scintilla."

Wm. H. Byford[2] in 1876 published an interesting case of Dropsy of the Amnion, and I. E. Taylor[3] read before the N. Y. Academy of Medicine an essay entitled Is Craniotomy, Cephalotripsy, or Cranioclasm preferable to Cæsarean Section in Pelves ranging from $1\frac{1}{2}$ to $2\frac{1}{2}$ inches?

The subject of inversion of the uterus has, during the last half century, attracted considerable attention in this country, and the valuable contributions of White, Noeggerath, Emmet, and others, to it, have been elsewhere noticed. The remarkable fact that a uterus for a long time inverted may, by an effort of nature, replace itself, has received due notice, and the evidence of American physicians has sustained that given of the fact by Spiegelberg, Leroux, De la Barre, Thatcher, Rendu, Shaw, Beaudelocque, and others of Europe.

It must be borne in mind that the possibility of this

---

[1] Amer. Journ. Obstet., April, 1876, p. 97.
[2] Chicago Med. Journ. and Exam., January, 1876, p. 1.
[3] Med. Record, March 15, 1876.

occurrence has been boldly denied by high authority, and that accumulation of evidence upon it is desirable. The case of spontaneous reposition recorded by De la Barre[1] was presented by him before the Academy of Surgery of Paris, and Beaudelocque was appointed a committee to examine into its authenticity. He reported that the account was "totally false," yet some years afterwards he himself met with the occurrence of a similar case which convinced him of his injustice to De la Barre.

Meigs[2] publishes three such cases; Jason Huckins,[3] of Maine, one; and Chestnut,[4] in 1876, records a most striking case, in which, after twelve years of inversion, a uterus was spontaneously replaced. In the last case no doubt as to the diagnosis can be admitted, for during its progress careful examinations were made by Byford, O'Ferral, and others. Indeed by the former a trial at replacement was practised, which lasted for about two hours.

In the May number of the *Obstetrical Journal of Great Britain and Ireland*, appears an excellent lecture upon Face Presentations, by Penrose; and in the January number of the *American Journal of Medical Sciences*, the report of a case of ovariotomy by Gaillard Thomas, in which four days after operation eight and a half ounces of milk were transfused into the patient's veins with good result. Thomas's procedure was based upon the experience of Hodder, of Toronto, Canada, and Joseph W. Howe, of New York. The former transfused milk on three occasions, and the latter on two. In no case did evil consequences result, and in two of Hodder's cases life seemed to be saved by the process. Thomas's case was the sixth on record, and the results were excellent. Since his publication J. W. Howe has experi-

---

[1] Archiv. Gén. de Méd., 1868, t. ii. p. 393.
[2] Obstetrics.   [3] Thomas' Dis. of Women, 4th ed. p. 431.
[4] Amer. Practitioner, May, 1876, p. 284.

mented on the subject with very unfavourable results. Transfusion of milk practised upon seven dogs, has in every case resulted in death, and in one man suffering from pulmonary consumption in the third stage death from coma occurred a few hours after the operation. The subject demands, and is certainly worthy of, full and careful investigation. It is difficult to reconcile the discrepancy of results which now attaches to it. At present the subject stands thus: up to 1875 six transfusions upon the living subject with no evil result, and with three instances of great benefit: during 1876 one transfusion upon man, and seven upon dogs, with fatal consequences in every case.

In the yearly contributions to medical literature there is a great deal of faithful, arduous, and useful work done, which redounds but little to the immediate advantage of the doers. This is the work done by reviewers. To J. C. Reeve, of Dayton, Ohio, this department of medicine is much indebted in this respect. His reviews of the subject of anæsthesia which have appeared in the *American Journal of the Medical Sciences*, are well known.

The mental development which appears thus far to have resulted from the peculiar education and training which characterize the civilization of this country, exhibits a much more marked tendency to the adaptation of means to immediate practical results, than to a devotion to abstruse study or pains-taking scientific investigation. Hence a fruitful harvest would naturally be expected in the way of ingenious appliances and well-conceived instruments, the outcome of a century of experimentation. This expectation will not be disappointed either in this or any other of the practical departments of medicine.

A vast number of modifications of the obstetric forceps, both short and long, have been made; so large a number, indeed, that even a mention of them would prove impossible. The most valuable and generally popular of these is the long forceps of Hodge. Two other excellent modifications are those of the late George T. Elliot and of J. P. White. The two latter are light, yet powerful; elegant in shape; and well adapted to the varied requirements of this most useful of surgical instruments.

Of vaginal specula there is rapidly being created as great a variety as that of forceps. Sims' great invention, developing an entirely new method of examination, certainly takes the lead of all others, and up to the present date none other can be compared with it for practical advantages. This instrument, however, requires two things for its employment—first, a certain degree of skill on the part of the operator in its use; and, second, an assistant to hold it during examination. To avoid the necessity of the second requirement, modifications have been made by Howard, Emmet, Hunter, Bozeman, Byrne, Nott, Otto, Noeggerath, and many others.

It would be useless to enumerate, as an original conception, each instrument employed in the operation for cure of vesico-vaginal fistula, for all of these were invented by Sims, as the pioneer in this procedure.

The uterine repositor of Sims is the best instrument yet devised for replacing the retroflexed or retroverted uterus. It is far superior to the ordinary uterine sound in efficiency, and unattended by its dangers. The same remarks apply to Sims' silver uterine probe, as compared with the unyielding sounds of Simpson, Huguier, and Kiwisch.

In operations upon the vagina and perineum, Emmet's curved scissors are very useful, and greatly facilitate these procedures; and after operations for atresia, Sims' vaginal plug of hard rubber or glass is indispensable.

For dilating a constricted uterine neck, Molesworth has furnished us an excellent instrument in his hydrostatic dilators, which, though acting upon the same principle as the water-bags of Dr. Barnes, are more powerful and manageable.

The syringe of Davidson is a valuable one for accomplishing vaginal irrigation, and the induction of premature delivery.

In many operations for the removal of abdominal tumours, temporary control of hemorrhage can be perfectly accomplished by H. R. Storer's clamp-shield, which becomes, under these circumstances, a valuable instrument. Permanent clamps have been devised by Atlee, Dawson, Thomas, and Greene, of Portland. The last of these consists of a spring clamp, intended to cause ligatures placed around the pedicle to cut through, and thus be liberated.

The galvano-caustic battery, only of late years introduced amongst us as a means of amputating vascular parts like the cervix uteri, etc., has now become very popular, and the ingenuity of Byrne and Dawson has furnished us with instruments at once small, portable, and very powerful. These instruments weigh only five or six pounds, and occupy little more space than an octavo volume. Their present dimensions and certainty of action remove two of the greatest objections attaching to the cumbrous and fickle instruments formerly in use.

For a long time after pessaries were put upon their proper basis as surgical appliances of great value, and as means which were essential to the proper management of uterine displacements, few modifications were made in them. Of late years, however, this has not been so. Hodge's instrument has been usefully modified by Albert H. Smith.[1] Many varieties of vaginal stem pessaries have been devised for prolapsus; and Ephraim Cutter, of Boston, has accomplished a valuable improvement in retroversion pessaries

---

[1] Obstet. Journ. of Great Britain, Amer. Sup., 1875, vol. iii., p. 7.

by getting support by a stem arching backwards over the perineum, and attaching to a belt worn around the waist.

In certain operations upon the anterior vaginal wall, the apparatus of Bozeman, by which the patient can be kept in a modified genu-pectoral attitude, proves very useful; by its use anæsthesia may be kept up for a long time with perfect comfort to the patient.

John T. Hodgen, of St. Louis, has made the needles employed in operations for vesico-vaginal fistula trocar-pointed, with great advantage. Their power of penetration is great, while at the same time they do little damage by cutting the tissues.

Parvin's polyptome is a very useful instrument for the removal of growths attached in utero, which are out of reach of manipulations practised by the instruments ordinarily in use.

A most valuable improvement in the trocar and canula for tapping the abdomen and abdominal tumours has been effected by S. Fitch, in his "dome trocar." By this instrument complete protection is given to the viscera by a projecting piece which shields them from its sharp point.

An excellent double canulated tube has been introduced for pelvic drainage by George H. Bixby, of Boston. It fulfils every requirement under these circumstances as to thoroughness and facility of employment.

LITERATURE AND INSTITUTIONS.

BY
JOHN S. BILLINGS, M.D.,
ASSISTANT SURGEON UNITED STATES ARMY.

LITERATURE AND INSTITUTIONS.

"Wherefore, by their fruits ye shall know them."

BESIDES his duties to his patients, the physician is under certain obligations to contribute, by way of interest, his quota to the common stock of medical knowledge from which he has drawn so freely. The skilful diagnosis, judicious medication, or bold and successful operation, if not properly recorded, benefit the individual only, not being available for those comparisons and higher generalizations which alone can make medicine a science. By the manner in which this duty, of preserving and transmitting the results of its labour and experience, has been performed, the medical profession of a country, as well as the individual physician, must to a great degree be judged, and the question now presented is, to what extent and in what manner have the physicians in the United States fulfilled this part of their professional obligations during the century just passed.

In the retrospective reviews, historical sketches, and centennial addresses which have, during the past year, been devoted to American medicine, our most important contributions to the healing art have been duly pointed out, and for the most part sufficiently eulogized. That the United States has a medical literature, has been cumulatively demonstrated, even to the extent of raising a suspicion of the existence of a doubt upon this point; and that this literature contains many valuable original contributions to the art, if not to the science, of medicine may be considered as unanimously affirmed and admitted.

If the defects of which all are more or less aware, have been but slightly referred to, it is because the purpose of the writers has been rather eulogistic than critical. In this final article of the present series, the object is not to select for praise the best of the work, nor the reverse, but to endeavour to give an idea of the quantity and value of the whole of it. So far as individual writers are concerned, an attempt will be made to supplement the information given in previous papers, but these have been so complete as regards that which is worthy of notice, that little need be said of single books and articles.

We will first endeavour to give some account of the quantity of medical literature produced in the United States during the last hundred years; making use for the purpose of some statistics obtained from a nearly complete list of the medical books published in this country from 1776 to the present time, and from which it may be considered certain that no important work has been omitted.

In these statistics we do not include works intended for the non-medical public, those relating to "ics" or "pathies," nor the great mass of what are called pamphlets in the technical sense of the word, that is, books of less than one hundred pages. The great majority of these pamphlets are either reprints from periodicals, addresses inaugural or valedictory, a few of which contain historical data of interest, or controversial and personal disquisitions which are best forgotten. While it is true that there is no necessary connection between the size of a work and its practical or scientific value, it will be found that with a very few exceptions, which have been pointed out in the preceding articles of this series, nothing of interest or importance is omitted by this division. The books to be counted may be classified as follows:—

I. Systematic treatises and monographs by physicians residing in this country, including reports of hospitals, corporations, and government departments.
II. Reprints and translations of foreign medical books.
III. Medical journals.
IV. Transactions of medical societies.

The first, third and fourth classes include what is ordinarily meant by the phrase "American Medical Literature." From them are excluded books written by American authors, but printed abroad, as, for instance, those of Dr. Wm. Charles Wells; while on the other hand, they include books written by physicians born and educated abroad, but who may be said to have become citizens of this country, such as Tytler, Pascalis, Bushe, Dunglison, Jacobi, and Knapp.

The statistics of the four classes above given, include not only the medical literature of the United States for the century, but nearly all which the country has produced since the first settlement. At the commencement of the Revolutionary War, we had one medical book by an American author, three reprints, and about twenty pamphlets. The book referred to is the "Plain, Precise, Practical Remarks on the Treatment of Wounds and Fractures," by Dr. John Jones, New York, 1775. It is simply a compilation from Ranby, Pott, and others, and contains but one original observation, viz., a case of trephining followed by hernia cerebri.

The libraries of our physicians were composed, according to Bartlett,[1] of the works of Boerhaave, with the Commentaries of Van Swieten, the Physiology of Haller, the Anatomy of Cowper, Keil, Douglass, Cheselden, Monroe, and Winslow; the Surgery of Heister, Sharp, Le Dran, and Pott; the Midwifery of Smellie; the Materia Medica of Lewis; and the works of Sydenham, Whytt, Mead,

---

[1] A Dissertation on the Progress of Medical Science in the Commonwealth of Massachusetts, Boston, 8vo., 1810.

Brookes, and Huxham. The works of Cullen were just beginning to be known. The only public medical library was that of the Pennsylvania Hospital, which contained, perhaps, two hundred and fifty volumes. There were probably not two hundred graduates of medicine in the country, and not over three hundred and fifty practitioners of medicine who had received a liberal education. Two medical schools had just begun, but had accomplished little previous to the war which closed them, there were no medical journals, and but one State Medical Society, that of New Jersey, had been organized. From this unpromising condition of things, have been developed the literary results, of which we now present a summary.

Table showing number of Medical Books printed in the United States from January 1, 1776, to January 1, 1876.

		1775 to 1799	1800 to 1809	1810 to 1819	1820 to 1829	1830 to 1839	1840 to 1849	1850 to 1859	1860 to 1869	1870 to 1875	Total
American Medical Books	CLASS I.										
	No. 1st edition	39	24	51	48	83	96	101	157	130	729
	No. later editions	9	4	14	17	34	49	80	85	44	336
	No. Vols. Total	51	31	77	86	136	162	197	256	180	1176
Reprints and Trans.	CLASS II.										
	No. 1st edition	28	39	64	72	145	135	99	104	81	767
	No. later editions	11	23	28	33	36	67	76	64	50	388
	No. Vols. Total	49	76	111	135	192	214	184	160	137	1274
Medical Journals	CLASS III.										
	No. Journs. com'ced	1	5	6	17	18	26	52	38	32	195
	" " discont'd	......	3	5	10	18	14	31	36	20	137
Original	"A."										
	No. Vols. com'nced	2	21	27	85	104	173	376	292	296	1376
	No. Vols. compl'ted	2	20	27	79	98	166	366	271	283	1312
Reprints	"B."										
	No. Journals	......	......	1	4	......	5	1	3	3	17
	No. Volumes	......	......	9	29	20	46	71	51	32	258
Transactions Med. Societies	CLASS IV.										
	No. Volumes	7	3	2	5	17	27	76	88	111	336

It will be seen from this table, that the medical literature of the United States really commences with the present century, and this is still more apparent, if the character of the works issued prior to 1800, be considered.

The first literary contributions of our physicians, after the close of the war, are contained in the memoirs of the American Academy of Arts and Sciences, Boston, 1785, and in the Transactions of the American Philosophical Society at Philadelphia, 1786. The first original separate work was the "Cases and Observations by the Medical Society of New Haven County, in the State of Connecticut," New Haven, 86 pp., 8vo., 1788. This is a collection of twenty-six articles, including several cases and autopsies, of interest, and a paper on the production of dysentery among troops by overcrowding and foul air, in which the connection of cause and effect is clearly demonstrated.

The majority of the succeeding publications, to the end of the century, related to the yellow fever, which was then epidemic along the whole Atlantic coast. The most prominent author of this period is Benjamin Rush, noteworthy also as an orator and politician. His writings excel in manner rather than matter, and the undoubted influence which he exerted over the earliest stages of American medicine, was probably due to his lectures rather than his published works. The best of his essays, and indeed the only one to-day worth consulting, is that on diseases of the mind, which contains some original observations of interest. One of his eulogists, Dr. Ramsay,[1] says: "On the correctness of this opinion [viz., his fondness for the use of the lancet] his fame as an improver of medicine in a great degree must eventually rest." And to the correctness of this judgment we entirely assent.

---

[1] Eulogium upon Benjamin Rush, by David Ramsay, Philadelphia, 1813, 8vo., pp. 79.

The work of James Tytler[1] is a good compilation, and contains, among other data not to be found elsewhere, an interesting letter by Dr. John Warren, of Boston. Tytler was born in Scotland in 1747, came to this country about 1796, and was drowned in 1804; he possessed extensive and varied learning, and wrote much, but for the most part on non-medical subjects.

The works of Noah Webster,[2] though mainly historical, are still of interest, and worth preservation.

Another writer of this period is Dr. William Curry, a native of Pennsylvania, 1755–1829. At first educated for the church, he acquired an excellent knowledge of Latin and Greek, and studied medicine under Dr. Kearsley, of Philadelphia. During the Revolutionary War he served as surgeon in the American army, being attached to the military hospital on Long Island, in 1776. After the war, he at first settled at Chester, but removed to Philadelphia about 1791. He was one of the original fellows of the College of Physicians of Philadelphia, and for many years a member of the Board of Health. His principal works in addition to his numerous pamphlets and articles on yellow fever, are his "Historical Account of the Climates and Diseases of the United States," 1792; and his "View of the Diseases most Prevalent in the United States," Philadelphia, 1811.

Towards the close of the century, and for a few years thereafter, there were published in Boston, New York, and Philadelphia, a number of medical theses, which, being

---

[1] A Treatise on the Plague and Yellow Fever, with an Appendix, 8vo., 1799.

[2] A Collection of Papers on the subject of Bilious Fevers, prevalent in the United States for a few years past. 246 pp. 8vo. New York, 1796. A Brief History of Epidemic and Pestilential Diseases; with the principal Phenomona of the Physical World which precede and accompany them, and Observations deduced from the facts stated. 2 vols., 8vo., Hartford, 1799.

classed as pamphlets, are not taken into account in our statistics, and are noticed here for the sake of saying a word with regard to this class of medical literature. A medical dissertation prepared, not for the press, but simply as a formality necessary for the obtaining of a diploma, as is the case with nearly all those which have been presented at our medical schools for the last fifty years, fairly merits the denunciation of Professor Gross, "that not one in fifty affords the slightest evidence of competency, proficiency, or ability, in the candidate for graduation."

Such was not the case, however, with regard to the theses above referred to, nor can it be justly said with regard to any series of printed theses of the European schools. It would seem, therefore, that when prepared as they should be, with reference to the probable criticisms, not merely of a single professor, but of the press and the public, there is the strongest inducement to refrain from plagiarism, and to produce the best work of which the candidate is capable; and it is well known to those who have had frequent occasion to consult them, that collections of printed medical theses are valuable, as historical documents, presenting a reflex of the teachings of the school, and as containing accounts of cases and original investigations, or particular doctrines of the student's preceptor, which cannot be found elsewhere. The proportion of copied matter, vague speculations, and other rubbish, does not, upon the whole, appear to be so much greater in this than in some other classes of medical literature, as to warrant their wholesale condemnation; and the remedy for the present unsatisfactory character of the theses of our medical students, appears not to be their abolition, but the requiring that they shall be printed, and considered as an important and real test of the merit of the candidate. They should of course be written in the vernacular. The influence which a teacher has in directing the thoughts of his pupils, is very well shown in the theses of the Philadelphia school, a considerable number of which related to medical botany, under the stimulus given by Dr. Barton to that branch of study.

During this period, and prior to the establishment of any medical journal, or regular publication of the transactions of any medical society, a number of communications from American physicians were sent to societies in Europe, and appear in their transactions. Perhaps the most notable paper of this kind was "An Experimental Inquiry into the Properties of Opium," by John Leigh of Virginia, which obtained the Harveian prize for 1785, and was printed at Edinburgh in the following year. It is worth consultation, not only for the facts which it records, but for the method of investigation pursued, which was unusual in that day of theories.[1]

From the year 1800 to the present time, the above table shows that there has been a steady increase in the amount of our indigenous medical literature, corresponding in the main to our increase in population and wealth. To obtain some notion of the quality and value of this production, a more detailed analysis is necessary.

The greater part of these books are compends relating to the treatment of diseases and injuries. Those which have been most popular, and are the best known, are the text-books and systematic treatises. These are for the most part compilations, but their importance is by no means to be underestimated, for the practice of the majority of the physicians of this country to-day, is based on the text-books of

---

[1] In this connection also may be mentioned a rare and little known work, being the oration delivered at the University of Virginia in 1782, by J. F. Coste, the Medical Director of the French Forces. Its subject is "Antiqua novum orbem decet medico philosophia;" it is dedicated to Washington, of whom the author was a personal friend, and makes a volume of 103 pages, 8vo.; printed at Leyden, in 1783.

the teachers in the New York and Philadelphia schools. Also we must remember that "there are compilations and compilations." The preparation of such systematic treatises as those of Flint, Gross, Stillé, and Wood, does not require less labour or thought, or give less scope for display of genius, than the so-called original monographs.

Writers of this class bring into their proper relations the isolated facts and observations scattered through many books, give them the mint stamp of value, and put them into general circulation.

For reasons already stated, and for want of space, but few books can here be noticed, even by title, and in connection with these will be given some very brief biographical data relating to a few authors. Of living writers and their works, as little as possible will be said.

In Anatomy our principal systematic works have been produced by Wistar, Horner, Morton, Richardson, Agnew, Hodges, Leidy, and Smith. None of them are now of interest. Dr. Caspar Wistar, 1761–1818, was of German descent, and a native of Philadelphia. Having obtained a good classical education, he studied medicine under Dr. John Redman, and took the degree of Bachelor of Medicine, in 1782. He continued his studies at Edinburgh, where he graduated M.D. in 1786. Returning to Philadelphia, he became Adjunct Professor of Anatomy in 1791, and continued to lecture until his death. His System of Anatomy was issued in parts, 1811–1814, making two volumes, and was a popular text-book for a long time.

The first work issued by Dr. Horner was a Dissector's Manual, in 1823. This was followed by his treatise on General and Special Anatomy in 1826, his Anatomical Atlas, and treatise on General and Special Histology.

A good original work has yet to be written on this last subject, in this country. In surgical anatomy, Drs. Anderson and Darrach have produced partial treatises, the first on the groin, pelvis, and perineum, New York, 1822; the second on the anatomy of the groin, Philadelphia, 1830.

Drs. N. R. Smith, Goddard, and Neill, have each issued a work on the Surgical Anatomy of the Arteries. Among the few original works in this department, should be mentioned those of Dr. John D. Godman, a native of Annapolis, Md., 1794-1830. Poor and almost friendless, but urged on by an unquenchable thirst for knowledge, he persisted in obtaining an education in spite of the greatest difficulties and discouragements, and at last took the degree of M.D. at the University of Maryland in 1818.

In 1821 he went to Cincinnati to accept a chair in the Medical College of Ohio, but dissensions in the faculty induced his speedy resignation. He then established a medical journal hereafter to be alluded to, but in 1822 went to Philadelphia and began a course of private lectures in anatomy. In 1826 he accepted the chair of Anatomy in Rutgers College in New York, but failing health soon compelled him to cease teaching, although he continued to use his pen until just before his death. Dr. Godman was an anatomist by nature, and though the necessities of breadwinning prevented him from accomplishing any great work, his treatise on the fascia[1] and his contributions to physiological and pathological anatomy[2] are really original and valuable productions.

The papers of Dr. John Dean on the "Microscopic Anatomy of the Lumbar Enlargement of the Spinal Cord," Cambridge, 1861, and on "The Gray Substance of the Medulla Oblongata," published by the Smithsonian Institution in 1864, are the results of careful work, and are note-

---

[1] Anatomical Investigations, comprising Descriptions of Various Fasciæ of the Human Body, 8vo., Philadelphia, 1824.

[2] Contributions to Physiological and Pathological Anatomy, 8vo., Philadelphia, 1825.

worthy for the use made of photo-lithography from microphotographs to obtain the illustrations.

The craniological works of Drs. Morton and J. A. Meigs should be referred to here. Dr. Samuel George Morton, 1799–1851, was a native of Philadelphia, and graduated in medicine at the University of Pennsylvania in 1820, after which he continued his studies for three years at Edinburgh, obtaining his degree in 1823. From 1839 to 1843 he was Professor of Anatomy in the Pennsylvania Medical College. His fame rests upon his "Crania Americana," Philadelphia, 1839, and his "Crania Egyptiaca," ibid., 1844; works which have a world-wide reputation, and whose value is permanent. His labours in this direction have been continued by Dr. J. Aitken Meigs, whose "Catalogue of Crania," Philadelphia, 1857, is well known to all who are interested in this subject.

In physiology, our text-books have been the works of Dunglison, Draper, Dalton, and Flint, all too well known to require more than a mere reference. The work of Professor Draper, published in 1853, was the first in this country in which micro-photographs were used to obtain illustrations. To these may be added the works of Reese, Oliver, Goadby, and Paine. Of special treatises and essays, the most important are Beaumont's Experiments on Digestion, Plattsburgh, 1833; Draper "On the Forces which produce the Organization of Plants," New York, 1844; Joseph Jones' "Investigations," published by the Smithsonian in 1856; S. W. Mitchell's "Researches upon the Venom of the Rattlesnake," *idem*, 1860; and Hammond's "Physiological Memoirs," Philadelphia, 1863. In this department Brown-Séquard may be claimed as an American author; some of his researches having been made, and the results first published in this country. Those who are familiar with the literature of thirty years ago will remember with a smile, the treatise of Emma Willard on the circulation of the blood, and the controversies to which it gave rise. The "Essays on the Secretory and the Excito-Secretory System of Nerves," by Dr. H. F. Campbell of Georgia, Philadelphia, 1857, should be remembered in this connection, as also the pamphlets of Dr. Dowler of New Orleans.

In the department of Materia Medica and Therapeutics, we have made a good record. In Medical Botany, the works of B. S. Barton and Jacob Bigelow deserve especial mention as works of permanent value. The "Illustrations of Medical Botany," edited by Dr. Carson, Philadelphia, 1847, containing one hundred plates, in folio, is a rare and costly work, a considerable part of the edition having been destroyed by fire.

The first systematic treatise on Materia Medica and Therapeutics, produced in this country, was that of Dr. Chapman, Philadelphia, 1817. This was followed by the works of Eberle, J. B. Beck, Dunglison, Harrison, G. B. Wood, T. D. Mitchell, Biddle, Stillé, Riley, and H. C. Wood, all of which have been, or are popular text-books in the schools.

The majority of these authors will be referred to under other sections, but of three, a few words may here be said. Dr. John P. Harrison was born in Louisville in 1796; studied under Dr. Chapman, and graduated in medicine in 1819. He was Professor of Materia Medica in the Cincinnati College from 1836 to 1839. In 1841 he accepted the same chair in the Medical College of Ohio, in 1847 was transferred to that of Theory and Practice, and died of cholera in 1849. He was one of the editors of the Western Journal of Medicine, and of the Western Lancet; published a collection of his essays in 1835, and his "Elements of Materia Medica and Therapeutics" in 1846.

The principal work on Materia Medica is the "United States Dispensatory" of Wood and Bache. Dr. Franklin Bache was born in Philadelphia in 1792, and died in 1864. Graduating as Bachelor of Arts in 1810, he studied under Dr. Rush, and obtained his medical degree in 1814. His tastes led him to the special study of chemistry, of which branch he was appointed professor in the Franklin Institute, in 1826. In 1841 he accepted the same chair in the Jefferson School. His principal work was in connection with the United States Pharmacopœia and the Dispensatory, which have made his name familiar to every physician in the United States. The first proposal to form a Pharmacopœia in this country was made to the College of Physicians of Philadelphia, in 1787, with the result of the appointment of a committee, which seems to have continued about ten years, but effected nothing. In 1808 a Pharmacopœia was published by the Massachusetts Medical Society, and in 1816 another was issued by the New York Hospital. Our present national Pharmacopœia originated in a plan submitted to the New York County Medical Society, in 1817, by Dr. Lyman Spalding. A leading part in the formation of the first edition, by the convention which met in Washington in 1820 for that purpose, was taken by the College of Physicians of Philadelphia, through its delegates, and more especially by Dr. Thomas T. Hewson; and in the subsequent revisions, Drs. Hewson, Bache, and Wood were the principal workers. The first revision, adopted in 1830, was entirely the production of these gentlemen, and was substantially a new work. The Dispensatory was projected by Drs. Wood and Bache as an exposition of the Pharmacopœia, and a means of making it more popular.

The exposition has, so far as our physicians are concerned, entirely overshadowed the text, and in a financial point of view, the Dispensatory is the most successful medical book ever published in this country.

Among writers on Materia Medica, distinguished in their day, may be mentioned Dr. William Tully, 1785–1859, who graduated at Yale in 1806, and attended medical lectures at Dartmouth College in 1808–9. He received the honorary degree of M.D. from Yale in 1819. In 1824, he was appointed Professor of Theory and Practice in the Castleton School, and in 1826 removed to Albany, forming a partnership with Dr. Alden March. In 1829, he accepted the chair of Materia Medica and Therapeutics at Yale, and removed to New Haven, but continued his lectures in Castleton until 1838. He ceased teaching in 1841. His principal works were the "Essays on Fevers," published with those of Dr. Miner 1823, a work which gave rise to much controversy, and was, upon the whole, not favourably received; a prize essay upon Sanguinaria, published in the American Medical Recorder in 1828; some papers in the Boston Medical Journal; and finally, his treatise entitled "Materia Medica, or Pharmacology and Therapeutics," Springfield, 1857–58, in two large volumes 8vo. This was published in numbers, was not a popular work, nor calculated for the use of a student, but shows great industry and learning in every page. Complete copies of it are not now easily obtained, although it cannot be said to be rare. His style is discursive, diffuse, and polysyllabic, and a decided effort is necessary to peruse his writings; but his knowledge of facts was minute and exact, and his last work is a mine of information, which is even now worth exploring by the curious.

In Surgery, our indigenous text-books have been produced by Dorsey, Gibson, S. D. Gross, Ashhurst, and Hamilton. On Operative Surgery we have the treatises of Pancoast, Piper, H. H. Smith, Stephen Smith, and Packard. The posthumous work of McClellan is not a systematic treatise, but a series of essays and cases, in which the description of Shock is especially noteworthy as being true to life. Of monographs, the most valuable are those by Professor Gross, on Wounds of the Intestines, 1838; on Diseases of the Bladder, 1851–55; on Foreign Bodies in the Air-passages, 1854 and 1862; and Diseases of the Bones and Joints, 1830; F. H. Hamilton on Fractures and Dislocations, 1860, fifth edition, 1875; Durkee and Bumstead on

Venereal; Van Buren and Keyes, and Gouley on the Urinary Organs; Bushe on Diseases of the Rectum; Carnochan on Congenital Dislocations of the Head of the Femur; H. J. Bigelow on the Mechanism of Dislocation and Fracture of the Hip; Ashhurst on Injuries of the Spine; Markoe on Diseases of the Bones; and Garretson's Oral Surgery. Specially valuable collections of cases, are the works of John C. Warren, on Tumors, Boston, 1837; and of J. Mason Warren; the pamphlets of Sayre on Orthopædic Surgery; N. R. Smith on Fractures of the Lower Extremity; and J. C. Nott, "Contributions to Bone and Nerve Surgery." As an example of careful statistical work, the treatise of R. M Hodges on "The Excision of Joints," Boston, 1861, is to be specially commended.

The treatise of Dr. Gross, on Wounds of the Intestines, above referred to, first appeared in the "Western Journal of Medicine;" it contains the results of numerous experiments and observations, and is of much practical value and interest. It is a rare book, and a copy of it may properly be considered a prize by the collector.

In Military Medicine and Surgery nothing of value was produced by the revolutionary war, the war of 1812, or the war with Mexico. This deficiency has been, to a great extent, made up by the number and value of works resulting from our late war.

The Medical and Surgical History of the War will be, when completed, the largest medical work ever produced in this country. The publications of the Sanitary Commission, including the works of Flint, Gould, and Lidell, contain valuable data. The manuals of military surgery have been written by Gross, Hamilton, Tripler, Blackman, Chisholm, and Warren. Other works which should be remembered in this connection are, Woodward on Camp Diseases, the statistical reports and circulars issued from the Surgeon General's Office, and the medical statistics of the Provost-

Marshal General's Office, compiled by Dr. Baxter, making two handsome quarto volumes, which are a most valuable addition to our knowledge of anthropometry and medical topography.

In the departments of Theory and Practice of Medicine, we have produced a fair amount of monographs and text-books, the most important of the latter class being those of Chapman, Eberle, G. B. Wood, and Flint. The following is a brief outline of the lives of a few who were our principal writers and teachers in this branch of medicine, but who now rest from their labours. Among them, there are few, who, in their day, had a more extended reputation, or were more popular than Dr. Nathaniel Chapman.

Born in Virginia in 1780, he received an excellent general education, became a pupil of Dr. Rush, with whom he was a favourite, graduated at the University in 1800, then spent three years in Europe, one as a pupil of Abernethy, and two at Edinburgh, and in 1813 was elected to the chair of Materia Medica in his Alma Mater, to be exchanged in 1816 for that of the Theory and Practice of Medicine, which he held until 1850, when he resigned. He died in 1853. His "Therapeutics and Materia Medica," published in 1817, was the best work of the kind in English at that date. He was the first President of the American Medical Association after its permanent organization; President of the American Philosophical Society, a popular lecturer, a genial companion, and in his prime probably the most distinguished physician in the United States. He edited, for seven years, the Philadelphia Journal of the Medical and Physical Sciences. Many of his lectures were published in the "Medical Examiner," in 1838-40. Two volumes of these lectures were published in 1844, and a compendium of his course on theory and practice was issued in 1846.

Contemporary with Dr. Chapman, and for twenty-five years associated with him as a teacher, was Dr. Samuel

Jackson, 1787-1872, a native of Philadelphia, and educated in the University of Pennsylvania, having graduated in medicine in 1808. From 1825 to 1863, he was Professor of the Institutes of Medicine in his Alma Mater. His "Principles of Medicine" (Philadelphia, 1832, 8vo.) was a treatise on pathology, founded on the doctrines of Broussais, and received high praise in its day. It was also the subject of a long and acrimonious critical review by Dr. Caldwell. The popular story that Dr Jackson recalled all the copies of this work that he could is incorrect; the entire edition was sold in the usual manner, and the publishers desired to issue another, but the author refused, on the ground that the science was undergoing such rapid and great changes that he would feel it necessary to re-write the entire work, a labour which his health and the demands of his private practice would not allow him to undertake. His most important writings are contained in the American Journal of the Medical Sciences, the last being a paper on a rare disease of the joints, in the July Number for 1870.

Dr. John Eberle, 1788-1838, was of German descent, and a native of Pennsylvania. After graduating in medicine in 1809, he went into politics, edited a newspaper, acquired intemperate habits, and became a bankrupt. Commencing life again, in 1825 he took the chair of Theory and Practice in the Jefferson School, which he held until 1831, when he removed to Cincinnati, and became connected with the Faculty of the Medical College of Ohio. In 1837, he removed to Lexington, Ky., to accept a chair in the Transylvania School, but could not lecture, and soon died. His treatise on the Practice of Medicine, first published in 1829, was, in its day, a very popular work, in part at least because of the formulæ which it contained, but is now forgotten.

Dr. Elisha Bartlett, born in Rhode Island in 1804, died 1855, graduated in medicine at Brown University in 1826, after which he spent a year in Paris. He held Professorships at Woodstock, Vt., Pittsfield, Mass., Dartmouth, Baltimore, Lexington, Louisville, and finally, in 1850, in the University of the City of New York. Of the numerous productions of his pen, the most noteworthy are the "Inquiry into the Degree of Certainty in Medicine," etc., Philadelphia, 1848; "The History, Diagnosis, and Treatment of Typhoid and Typhus Fever," Philadelphia, 1842; and, "The History, Diagnosis, and Treatment of the Fevers of the United States," Philadelphia, 1847; of which, three subsequent editions were issued. To these may be added his essay on the Philosophy of Medical Science, in which the importance of facts and observations is insisted on, and all theorizing is denounced, in accordance with the teachings of Louis.

Dr. David Hosack, 1769–1835, a native of New York, graduated as Bachelor of Arts at Princeton in 1789, and as Doctor of Medicine in the University of Pennsylvania in 1791. After practising a year at Alexandria, Va., he spent two years in Edinburgh and London. Returning to New York, he entered into partnership with Dr. Samuel Bard, was appointed Professor of Botany in Columbia College in 1795, to which was added the chair of Materia Medica, in 1797. In 1807, he was chosen Professor of Surgery and Midwifery in the newly-formed College of Physicians and Surgeons of the State of New York, and in 1813, took the chair of Theory and Practice. In 1826, he resigned, with others, and went into the Rutgers Medical College. His writings appear in the philosophical transactions, in the "Medical and Philosophical Register," of which he was the founder, and as occasional lectures and pamphlets. They were collected and published as "Essays on various subjects in Medical Science," in three volumes, New York, 1824–1830. His "System of Nosology" reached two editions; his "Lectures on Theory and Practice" were edited by Dr.

Ducachet, and published at Philadelphia in 1838. His most important paper was his "Observations on Febrile Contagion," and on the means of improving the Medical Police of the City of New York, N. Y., 1820. As a lecturer, editor, and writer, he exercised much influence on the profession, and his literary and scholarly tastes were imparted to his pupils, and especially to Dr. John W. Francis, who, after his graduation in 1810, became associated with him in practice. Dr Francis was the son of a German grocer, born in New York, 1789, died 1861. He was for thirteen years Professor in the College of Physicians and Surgeons, and followed Dr. Hosack to Rutgers, the close of which ended his career as a teacher.

Dr. Joseph Mather Smith, 1789–1866, graduated at the College of Physicians and Surgeons in 1815, and was Professor of Theory and Practice of Physic, same school, 1826 to 1855, when he took the chair of Materia Medica. He contributed largely to literature through the medical journals; presented some interesting reports to the American Medical Association, and published "Elements of Etiology," a "Philosophy of Epidemics," New York, 223 pages 8vo.

For beauty of style as a writer and lecturer, Dr. Samuel Henry Dickson is pre-eminent. Born in Charleston in 1798, he graduated at Yale in 1814, and in Medicine at the University of Pennsylvania in 1819; was Professor in the Charleston Medical School from 1824 to 1831, 1833–34, 1850–7; in the New York University, 1847–50; and in the Jefferson School, 1858; he died March 31, 1872. His systematic works were not very successful, or worthy of special remark, but his journal contributions, and especially his volumes of essays, are among the most attractive literature of medicine.

John K. Mitchell, born in Virginia, 1793, took his academical degrees at the University of Edinburgh, commenced his medical studies under Dr. Chapman in 1816, and graduated in medicine at the University of Pennsylvania in 1819. After three voyages to India and China, for the sake of his health, he returned to Philadelphia, and in 1822, began to deliver lectures on Medical Chemistry in the Summer School. In 1841, he was elected to the chair of the Practice of Medicine in the Jefferson Medical College, which he filled to the date of his death in 1858. As an original investigator, and clear logical reasoner, his name stands among the highest, and is probably destined to a higher relative position in the future, than it enjoys even now. His papers on Endosmosis, Mesmerism, Ligature of Limbs for Spasm, and Cryptogamous Origin of Fevers, will be consulted, not only for the original facts which they set forth, but as models of suggestiveness, if the phrase may be permitted.

Dr. Charles Frick, born at Baltimore August 8, 1823, received the degree of Doctor of Medicine from the University of Maryland in 1845. In 1856, he was chosen to the chair of Materia Medica in the Maryland College of Pharmacy, and in 1858, he became Professor of Materia Medica and Therapeutics in the University of Maryland. His most valuable contributions to literature are his "Analysis of the Blood," American Journal of the Medical Sciences, January, 1848; "Treatise on Renal Diseases," 1850; "On Diabetes," American Journal of the Medical Sciences, 1852; "On Urinary Calculi," American Medical Monthly, April, 1858. He died March 25, 1860, of Diphtheria, contracted from a patient upon whom he had performed the operation of Tracheotomy five days previous. All his papers are careful, conscientious reports of original observations, with the least possible amount of theory, and with direct reference to practice.

Among the diseases which have received the greatest amount of attention in this country may be mentioned yellow and malarial fevers, and diseases of the chest. Our literature on yellow fever includes over one hundred books and pamphlets, besides more than six hundred journal articles. It was the epidemic of this disease along the North Atlantic coast which gave the first impetus to medical authorship in this country, and produced a mass of controversial writings which, although of little value in a scientific point of view, were useful, as giving their authors the habit of writing for the press. The earlier books have already been referred to, but mention should be made of the writings of Felix Pascalis Ouviere, generally known under the name of Pascalis. Dr. Pascalis was a native of Provence, France, and was born about 1750. Having graduated in medicine at Montpellier, he went to St. Domingo, and there practised his profession until driven out by the Revolution of 1793, when he came to Philadelphia, and subsequently settled in New York, where he died in 1833. Besides his works on Yellow Fever, he wrote a treatise on Syphilis, New York, 1812, and contributed papers to journals. He was one of the editors of the Medical Repository.

Another writer on Yellow Fever who seems to be little known except in the South is Dr. J. L. E. W. Shecut, a native of South Carolina, born in Beaufort, 1770; died in Charleston, 1836. He studied under Dr. Ramsay, of Charleston, graduated M.D. at Philadelphia in 1791, and at once commenced practice in Charleston.[1] His most important essays were collected and published in one volume, Charleston, 1819, under the title of "Shecut's Medical and Philosophical Essays." This book, which is quite rare, contains his account of the yellow fever of 1817, first published in that year, and also his "Essays on Contagions and Infections," first published in 1818, and should be consulted by those who wish to trace the history of opinions in the South relating to this disease.

---

[1] For these data I am indebted to Dr. Robert Lebby, of Charleston.

The principal work on Yellow Fever, which includes the information of all others of a prior date, is that of Dr. Réne la Roche, published in 1855. Dr. La Roche was of French descent, born in Philadelphia, in 1795, his father being an emigrant from St. Domingo. Unlike the majority of prominent American physicians, he was not connected with a large medical school, and his justly deserved reputation rests entirely upon his writings, and especially on his treatise on Yellow Fever, which is a model of research, and is remarkable, not only for the number, but the accuracy of its references, and the impartiality with which opposing statements are given.

The most valuable recent articles on this disease are in the New Orleans and the Charleston Medical Journals, but the great majority of them are historical and controversial.

During the course of an epidemic, physicians are too busy to make observations which require much time or care, or to make more than brief notes. The papers of Drs. Faget,[1] Logan,[2] and Sternberg,[3] giving temperature observations, make an advance in the right direction, but we lack data as to the pathological chemistry of the disease, and as to its relations with the malarial fevers. With regard to this last class of diseases, our literature is even more extensive than that of the preceding, and occupies much space in the journals of the West and South.

Our most valuable contribution to the natural history of malarial disease is the treatise of Dr. Daniel Drake, on the principal diseases of the Interior Valley of North America. This work is the "Magnum Opus," and results of the life-long labour, including extensive personal observations, literary research, and matured reflection, of a man whose

---

[1] New Orleans Med. and Surg. Journal, 1873, i., N. S., p. 145.
[2]   Do.   do.   1874, ii. p. 779.
[3] Amer. Jour. Medical Sciences, 1875, lxx. p. 99.

fame, as compared with that of his contemporaries, will probably be greater a century hence than it is to-day, and whose name, even now, should be among the first on the list of the illustrious dead of the medical profession of the United States. The son of an illiterate Kentucky pioneer, brought up in a log cabin, attending a country school in the winter, and using the remainder of the year working on a farm, he surmounted the obstacles thus placed in his way, and by unceasing labour, joined to a sound common sense, which rose to the level of genius, took a leading position as author, editor, practitioner, and teacher. Commencing the study of medicine at the age of sixteen, he attended his first course of lectures in 1805, and his second in the University of Pennsylvania, in 1815, at the end of which he graduated. He was Professor successively in the Transylvania School, the Medical College of Ohio; a second time in the Transylvania; the Jefferson School; the Medical Department of Cincinnati College; the University at Louisville; and again in the Medical College of Ohio. He died November 6, 1852. His first publication was a pamphlet on the climate and diseases of Cincinnati, published in 1810, and reissued as "The Natural and Statistical View or Picture of Cincinnati and the Miami Country," published in 1815. This work is quite rare, and is interesting as being the germ from which sprung his great work above referred to.

He founded the "Western Journal of the Medical and Physical Sciences," which would be of much value, if for no other reason, on account of a series of essays on Medical Education, by Dr. Drake, which were published in it. These essays were issued in a separate volume, in 1832, and form upon the whole, the most satisfactory contribution to this vexed question which this country has ever produced. He commenced the preparation of his work on the diseases of the Mississippi Valley in 1822, and the second volume was not issued until after his death. Very few of the younger physicians of this country are familiar with his writings. Of his essays on Medical Education and Diseases of North America, no second editions have been published; but if there are any books to which the hackneyed phrase of the reviewer, "No physician's library is complete without it," apply, it is to these works of Dr. Drake, as far as American physicians are concerned, and they are most distinctively and peculiarly American books, in subject, mode of treatment, and style of composition.

The dissertation of Dr. J. K. Mitchell "On the Cryptogamous Origin of Malarious and Epidemic Fevers," is an ingenious piece of reasoning, and presents a summary of all the *à priori* arguments in favour of this theory which can be advanced. The papers of Dr. Salisbury on the same subject are without value.

Upon the subject of diseases of the chest the most noteworthy monographs have been the works of Morton, McDowell, Lawson, and Flint on Consumption; of Horace Green on the Diseases of the Air-passages; La Roche on Pneumonia, and of Gerhard and Flint on Diagnosis of Diseases of the Chest. The treatise on Phthisis, by Dr. L. M. Lawson, adds another to the numerous examples of careful studies by physicians of diseases with which they are themselves afflicted. Dr. Lawson was a native of Kentucky; born 1812, died 1864. His early education was defective. At the age of twenty he was licensed to practise, but it was not until 1838 that he obtained his diploma from the Transylvania School. In 1844 he was elected to a Professorship at Lexington; from 1847 to 1853 he filled the chair of Materia Medica in the Medical College of Ohio, and then became Professor of Principles and Practice of Medicine. During the winter of 1859–60, he lectured on Clinical Medicine in the University of New Orleans. He founded, and for a long time conducted, the "Western Lancet," in which many of his lectures were published.

Dr. W. W. Gerhard, 1809–72, was a native of Philadelphia, and a graduate of the University of Pennsylvania. After taking his degree he spent two years in Paris, and became thoroughly indoctrinated with the teachings of Louis. On his return to Philadelphia he was appointed lecturer at the Medical Institute, and Assistant Clinical Lecturer to Professor Jackson. For twenty-five years he was the senior Physician to the Pennsylvania Hospital. Some of his clinical lectures appeared in the "Medical Examiner," of which he was one of the editors. His principal work was his "Treatise on Diagnosis of Diseases of the Chest," Philadelphia, 1842; second edition, 1846.

Dr. Horace Green, 1802–1866, was a native of Vermont, and a graduate of Castleton Medical College in 1824. From 1840 to 1843 he was Professor of Theory and Practice in the same school; and in 1850 took the same chair in the New York Medical College, of which he was one of the founders, continuing to lecture until 1860. In connection with this school he established, with his colleagues, the "American Medical Monthly." He was the first in this country to devote himself to a specialty, and his works on the local treatment of diseases of the air-passages attracted much attention, although they are not of a character to add permanently to his fame.

In medical jurisprudence, the systematic works of Beck, and Wharton and Stillé, and the treatise of Dr. Wormley on Poisons, are the most important, and each of them compares most favourably with any similar works in existence.

There are probably not to be found in the annals of medicine so large and valuable contributions to its literature by three brothers, as were made by the Beck family of New York.

John B. Beck, 1794–1851, graduated in Columbian College in 1813, became a pupil of Dr. Hosack, and graduated in Medicine at the College of Physicians and Surgeons in

1817, presenting, as a thesis, a paper on Infanticide, which was published, and is still a standard work on this subject. In 1822 he assisted in establishing the "New York Medical and Physical Journal," with which he was connected for the next seven years, and in which he published numerous articles. In 1826 he became Professor of Materia Medica in the College of Physicians and Surgeons, just newly organized. His principal works, in addition to those already alluded to, were his "Essays on Infant Therapeutics," New York, 1849; second edition, 1855; and his Historical Sketch of the State of Medicine in the American Colonies; "Lectures on Materia Medica," and a collection entitled, "Researches in Medicine and Medical Jurisprudence."

Theodoric Romeyn Beck, 1791–1855, graduated at Union College, Schenectady, studied under Dr. Hosack, and graduated as M.D. at the College of Physicians and Surgeons, in 1811. He was appointed Professor of the Institutes of Medicine and Medical Jurisprudence in the College at Fairfield, in 1815. In 1817 he became Principal in the Albany Academy, and gave up the practice of medicine. In 1840 he took the chair of Materia Medica in the Albany Medical College, which he held until 1854. His great work was his treatise on Medical Jurisprudence, which appeared in 1823, in two volumes, and of which, including four English editions, ten editions were issued during the author's life.

Dr. Lewis C. Beck, 1798–1853, the younger brother of the preceding, studied medicine under Dr. Dunlop, and was admitted to practise in 1818. In 1826 he was elected Professor of Botany and Chemistry in the Vermont Academy of Medicine. This position he resigned in 1832. In 1836 he was appointed Mineralogist to the Geological Survey of the State of New York, and in 1840 was elected Professor of Chemistry and Pharmacy in Albany Medical College. His contributions to medical literature, to chemistry, meteorology, and mineralogy, were numerous. His principal medical work was his Report on Cholera, made to the Governor of New York in 1832.

The literature of obstetrics has been so fully given by Dr. Thomas, in a preceding article of this series, that further reference to it is superfluous. We will add only, with regard to Dr. Hugh L. Hodge, that he was a graduate of Princeton, a pupil of Dr. Wistar, and that his early taste was for surgery rather than obstetrics. He was induced to change his specialty by Dr. Dewees. He was afflicted with defective vision, which increased with age, and his great work on Obstetrics was produced entirely by dictation. He commenced as a lecturer in the Medical Institute, and was elected Professor of Obstetrics in the University of Pennsylvania in 1835, the rival candidate being Dr. Charles D. Meigs, a lecturer in the Philadelphia Association for Medical Instruction, who six years later obtained the chair of Obstetrics in the Jefferson School. The literary works of Dr. Meigs compare very unfavourably with those of his rival as to scientific value and exactness, but they are much more attractive to students and those who read for pleasure rather than instruction.

We have three names of American medical writers whose works should be mentioned here, viz., Coxe, Watson, and Dunglison.

Dr John Redman Coxe, 1773–1864, was a type of the medical scholar, who loves books for their own sake, and who takes more pleasure in discovering a forgotten sentence in a folio of the fifteenth century than in original investigations in the light of the present day. Born in Trenton, New Jersey, he completed his classical education at Edinburgh, studied medicine under Dr. Rush, and took his degree of M.D. at the University of Pennsylvania in 1794, after which he continued his medical studies in London, Edinburgh, and Paris for about two years. He was elected Professor of Chemistry in the University of Pennsylvania in

1809, and of Materia Medica and Pharmacy in 1818. He filled the latter chair until 1835, at which date he retired, and was but little known thereafter. His Dispensatory[1] and Medical Dictionary[2] were useful compilations, and met an existing want. His Observations on Vaccination[3] was his best original contribution to medicine. His Inquiry on the Discovery of the Circulation of the Blood was a paradoxical attempt to disprove the claims of Harvey. His last work, and the one most in accordance with his tastes, was "The Writings of Hippocrates and Galen," Philadelphia, 1846. He founded the first medical journal published at Philadelphia, preceding that published by Dr. Benj. Smith Barton by two months, and his library was, in its day, the best collection of ancient authors on medicine in this country.

Dr. John Watson, of New York, has been alluded to in the article on surgery. His literary tastes led him to historical studies and the collection of a valuable library, and his historical sketch of ancient medicine[4] shows that he consulted and enjoyed consulting the original works of the fathers in medicine.

Dr. Robley Dunglison, a native of Keswick, England, born in 1798, was one of the most prolific of medical authors. He obtained his medical education at Edinburgh, Paris, and London; settled in the latter city, where he wrote a treatise on the diseases of children [1824], and was one of the editors of the London Medical Repository in 1823–24. In 1824 he accepted the invitation of Thomas Jefferson to fill the chair of Anatomy, Physiology, Materia Medica, and Pharmacy, in the University of Virginia. At this place he published in 1827 a syllabus of his course on Medical Juris-

---

[1] The American Dispensatory. Phila., 1806, 4th ed. 1818.
[2] The Philadelphia Medical Dictionary. Phila., 1808, 2d ed. 1817.
[3] Practical Observations on Vaccination. Phila., 1802.
[4] The Medical Profession in Ancient Times. 8vo., N. Y., 1856.

prudence and prepared his Medical Dictionary. In 1833 he took the chairs of Materia Medica, Therapeutics, Hygiene, and Medical Jurisprudence in the University of Maryland, and from 1836 to 1868 was Professor of the Institutes of Medicine in the Jefferson School. He died April 1, 1869. His Systems of Physiology (first edition 1832), Hygiene (first edition 1835), Therapeutics (1836), Practice (1842), and Materia Medica (1843), were popular in their day, nearly all of them passing through several editions. The work by which he will be remembered is his Medical Dictionary. The first edition of this was published at Boston in 1833, in two volumes. A peculiarity of this edition is that it contains brief biographical sketches of physicians, omitted in subsequent issues. The last edition, Philadelphia, 1874, edited by his son, is the most convenient work of the kind in existence.

Our literature on insanity and the pathology of mental disease is insignificant in comparison with the importance of the subject and the opportunities existing for its study, the only monograph of permanent value being the "Contributions to Mental Pathology," by Dr. Isaac Ray, 8vo., Boston, 1873. Considering the number and size of the asylums for the insane in this country, and the amount of money which has been spent upon them, it is rather curious that the medical officers connected with them should have contributed so little to the diagnosis, pathology, or therapeutics of diseases of the nervous system. An examination of the works relating to this subject, and more especially of the American Journal of Insanity, which is the most important, and which contains the transactions of the Association of American Superintendents of Hospitals for the Insane, will show that the thoughts of these specialists have been mainly directed to the subjects of construction and management of asylums and to the jurisprudence of insanity. This last subject is one of great and increasing importance;

but our contributions to its literature consist rather of opinions and ontological speculations than of scientific observations. The annual reports of our insane asylums consist, for the most part, of business and financial statistics, and are intended for the use of appropriation committees rather than of physicians. There are some signs, however, that more attention will hereafter be given to recording of the physical phenomena of mental disease, and it is to be hoped that we may soon have some published results from the pathological department of the Utica Asylum, which will stimulate other institutions to undertake similar work. No more promising field to-day exists in medical science for valuable discoveries than in the wards and laboratory of a large, well-appointed hospital for the insane.

Upon the subject of hygiene no systematic work has yet been produced in this country, with exception of the treatise on Military Hygiene, by Dr. Hammond. One of the principal writers in this department was Dr. John Bell, a native of Ireland, 1796–1875. He came to this country with his parents, who settled in Virginia in 1810, and graduated in medicine in the University of Pennsylvania, after which he lectured for some years in the Philadelphia Medical Institute, and for two years in the Medical College of Ohio. His treatise on Baths and Mineral Waters is the only comprehensive and respectable treatise on this subject published in this country. The most important contributions to the literature of hygiene which we have produced are the reports of the various State and municipal boards of health, most of which, however, are of comparatively recent origin, and it is to be hoped are only just fairly commencing their career of usefulness.

The subject of hospital construction and hospital hygiene has been much discussed in this country, the latest production being a large and handsomely illustrated work published by the trustees of the Johns Hopkins Hospital of Baltimore.

The publications of our municipal, State, and national governments, relating to vital and medical statistics, are among our most valuable contributions to medical literature. The reports of city and State boards of health show each year evidences of more careful investigation into the probable causes of disease and the means of removing or diminishing them, and the necessity and economic value of such work is slowly but steadily becoming apparent to the educated classes of the community by means of the publications referred to.

The circulars and reports of the medical department of the army are sufficiently well known, and within the last few years a series of reports have been commenced by the Medical Department of the Navy and by the Marine Hospital Service of the Treasury Department, which it is to be hoped will become important additions to our medical literature, not only in regard to statistics, but in the departments of hygiene, pathology, and therapeutics. It should not be forgotten by the physicians of the United States that they are, to a certain extent, responsible for the condition of the medical departments of the government, since the sympathy and opinions, expressed or implied, of the medical profession at large as to the work which these departments have done, or are trying to do, furnish the encouragement and stimulus which are necessary to the continuous production of good results, and also influence to a considerable extent the action of our legislators with regard to the officers of these departments.

The reports of the Surgeon-Generals of the Army, the Navy, and the Marine Hospital Service, while ostensibly presented to the Secretaries of War, the Navy, and the Treasury, are really, in a sense, made to the physicians of the country, who are the only competent judges as to whether the work is satisfactory, and commensurate with the means which have been allowed for its performance.

Of encyclopedic works, the result of the combined labour of many authors, like the great French dictionaries, but one specimen has been attempted in this country. This was the American Cyclopædia of Practical Medicine and Surgery, edited by Dr. Isaac Hays, of which two volumes, completing the letter "A" were published at Philadelphia in 1834-36, and reissued with a new title, "Medical and Surgical Essays," in 1841. The time is perhaps not far distant when a first-class publication of this character will be sufficiently in request in this country to warrant an attempt at its production.

Reprints and Translations.—The second class of medical works referred to in our statistics, includes the reprints and translations, which cannot be overlooked in an account of our medical literature, since they have formed an important part of the libraries of American physicians, even if quantity only be considered.

Prior to the Declaration of Independence, the largest and most important medical book printed in this country was the "Lectures on Materia Medica," of Cullen, issued at Philadelphia in 1775, in 4to., and advertised as "The very cream of physic," and as "absolutely necessary for all American physicians who wish to arrive at the top of their profession."[1]

In 1776 was published, at Philadelphia, the treatise of Van Swieten on the Diseases Incident to Armies, with Ranby on Gunshot Wounds, and Northcote on Naval Surgery, forming a small volume of 164 pages, which is usually found bound with the second edition of John Jones' "Practical Remarks," etc., of the same date, and was probably the principal guide of the army surgeons during the Revolu-

---

[1] A copy of this work was purchased by the Library of the Pennsylvania Hospital 1780, for £133 5s. currency, equal to £1 15s. specie.

tionary War. Cullen's "First Lines of the Practice of Physic"[1] was reprinted from a smuggled copy, in 1781, at Philadelphia, in two volumes, 8vo., and five later American editions, the last edited, with a great flourish of trumpets, by Dr. Caldwell, in 1822, attest its popularity.

For thirty years after the Declaration of Independence, the majority of the reprints were works of English and Scotch writers, and especially of the Edinburgh school, the favourite authors being Cullen, Brown, John Hunter, Benjamin Bell, Denman, Smellie, Hamilton, Beddoes, and Robert Jackson. The largest edition sold was probably of the "Edinburgh New Dispensatory." The only translations of French or German medical works issued in this country prior to 1800 were, Swediaur on Venereal, New York, 1788, and Blumenbach's "Elements of Physiology," Philadelphia, 1795. The first medical book printed in Louisiana was "Médicaments et précis de la Methode de M. Masdevall," a pamphlet of 48 pages, relating to the yellow fever, issued in 1796.

The beginning of the influence of the French schools, which for the next fifty years was so powerful in the United States, especially in surgery, is marked by the editions of Boyer and Desault, Philadelphia, 1805, to which rapidly succeeded the works of Alibert, Richerand, and Bichat. In this connection may be permitted a reference to two works which are omitted from our statistics, since they were intended for non-professional use, but which had an extensive sale, and indirectly exerted a very considerable influence, viz., Buchan's Domestic Medicine, of which several editions were issued, the most important being that of Philadelphia, 1795, revised by Dr. S. P. Griffitts, and the "Primitive

[1] A copy of this work was purchased by the Library of the Pennsylvania Hospital, 1780, for £135 5s. currency, equal to £1 15s. specie.

Physic," of John Wesley, of which there are several American editions of the last century.

Many foreign medical works have been issued in this country in connection with periodicals, such as the " Register and Library of Medical and Chirurgical Science," published at Washington, D. C., 1833-36, in which were issued "Bell on the Nerves," "Lawrence on the Eye," Velpeau's Surgery, etc.; The Select Medical Library, edited by John Bell; the American Medical Library, published under the supervision of Dr. Dunglison; and the "Medical News and Library," in which some valuable books have been issued.

The number of translations of French medical works which have been published in this country is one hundred and forty-eight (148). One hundred and one of these were issued prior to 1842, and only eight have appeared within the last ten (10) years.

The number of translations of German works issued has been sixty-four (64), of which but fourteen (14) were issued prior to 1842, and twenty-eight (28) within the last ten years.

The number of reprints of English medical books has been five hundred and eighty-four (584), thirty (30) of these were issued prior to 1800; two hundred and seventeen (217) during the next forty years, and three hundred and thirty-seven (337) since 1840, the production gradually increasing.

It is largely to French and German sources that we owe our works on pathology, pathological anatomy, pathological chemistry, and physiology.

The best systematic treatise on the practice of medicine from the German, published in this country, was that of Niemeyer, in 1869, the name of the author having been made somewhat familiar to the American public by a translation of his lectures on Phthisis, published the year previous. The works of Billroth on General Surgical Pathology, New York, 1871, Rindfleisch, a Text-book of

Pathological Histology, Philadelphia, 1872, are the books which are to-day directing the work of the younger professional men of the country. The Cyclopædia of the Practice of Medicine, edited by Ziemssen, now in course of publication, is the most extensive medical work, native or foreign, which has ever been issued in the United States, and is probably destined to exercise great influence upon our investigation of diseases, whatever it may do for the practice.

Of the translations from the French, the most important have been those relating to anatomy, physiology, and surgery. The favourite authors have been Bichat, D. J. Larrey, Boyer, Orfila, Magendie, Laennec, Cazenave, Baudelocque, Louis, Velpeau, Broussais, Cazeaux, Colombat, Ricord, Vidal, and Malgaigne.

It would be useless to give lists of the titles of these; it is sufficient to say that they include nearly every important monograph or text-book produced by English writers: from Cullen, Brown, and Darwin, to Bennett, Watson, and Aitken; from John Hunter, Benjamin, John and Charles Bell, Pott, Hey, and the Coopers, to Erichsen, Paget, and Holmes; and from Hamilton and Smellie to Simpson, Barnes, and Duncan. The works of nearly all the great English teachers have been quickly reproduced on this side of the water, and their modes of treatment are those followed by the majority of our practitioners.

A few medical books have been printed in Spanish at Philadelphia, for the Mexican trade, including the "Compendio de la Medecina," by J. M. Venegas, 1827. The number of reprints in this country has been largely due to the want of an international copyright law, for which reason publishers found it much cheaper to take the work of an English author gratis, than to pay an American writer for his MS. Sometimes the name of an American physician is given as editor of the reprint, but in most cases, this means little more than that he approves the book, the so-called editing being imperceptible. To this remark a few honourable exceptions should be made, such as the additions by John Bell to the lectures of Stokes, of Gerhard to Graves; the reprints of Copland's Dictionary, in which the bibliographical additions, made by Dr. Charles A. Lee, are numerous and valuable, the editions of Velpeau's Surgery by Mott and Blackman, and the editions of Aitken's Practice by Dr. Clymer, who has added much to the completeness of the work.

This so-called editing was the subject of some caustic criticism, and has of late years almost entirely disappeared. With regard to the merits of the International Copyright question, there has been much discussion. On the one side, it is truly said that the desire for books increases by the supply, and that the sale of the cheap reprints produces a market for indigenous productions. On the other side, it is affirmed with equal truth, that it deprives our own writers, to a great extent, of pecuniary inducements to labour. The question is one to be decided, however, by the laws of morality rather than expediency, and the majority of educated non-interested parties agree that the passage of an international copyright law would be an act in accordance with the dictates of common honesty and justice.

Undoubtedly, the cheapness and abundance of these republications have done much to diffuse knowledge among our practitioners, and the libraries of many physicians have been mainly composed of the "pepper and salt sheepskin covered Philadelphia reprints." Of late years there has been a marked improvement in the quality of paper and typography of our medical books, while the stout bindings of sheep and calf of fifty years ago, have been largely superseded by the more showy, but, at the same time, more flimsy cloth bindings now in vogue. The German fashion of publication in parts has been almost unknown, except as connected with periodicals, and it is to be hoped that it may be long before the annoyance and confusion which attend the Lieferung and Hefte may be connected with our medical publications. "The American Clinical Lectures," edited by E. C. Seguin, and published by G. P. Putnam & Sons, look in this direction most unpromisingly, and the publication of such totally unconnected papers, in a series of continuous paging, even if special paging is added, must be unhesitatingly condemned by all who have occasion to either make or to verify bibliographical references to them.

It may be of interest to refer to some statistics of the locality of publication of these works. Of class one (I) we find that three hundred and seventy-three (373) first editions were published in Philadelphia, one hundred and seventy-three (173) in New York, eighty-one (81) in Boston, twenty-four (24) in Cincinnati, sixteen (16) in New Orleans, and fifteen (15) in Baltimore, leaving ninety-six (96) published elsewhere. If each edition be reckoned as a separate work, we find that six hundred and thirteen (613) have appeared in Philadelphia, two hundred and twenty-six (226) in New York, ninety-six (96) in Boston, and eighteen (18) in Baltimore. Of the reprints and translations, six hundred and eighteen (618) books, or seven hundred and fifty-three (753) editions have been issued from Philadelphia, one hundred and seventy-seven (177) books, or two hundred and nineteen (219) editions from New York, eighty (80) from Boston, and ninety-four (94) elsewhere. It appears then that more than one-half of our medical books have been published in Philadelphia, and about one-fifth in New York. The firm of Carey, Lea & Carey, now H. C. Lea, has published nearly six hundred editions of medical works; and those of Lindsay & Blakiston, and Lippincott, each between one and two hundred. In New York, the principal publishing house is that of S. S. & W. Wood, now Wm. Wood & Co., which has issued about one hundred and fifty (150) editions.[1]

Medical Journals. — It is not in text-books or systematic treatises on special subjects that the greater part of the original contributions to the literature of medicine have been first made public during the last century, either in this or other countries. Since the year 1800 medical journalism has become the principal means of recording and communicating the observations and ideas of those engaged in the practice of medicine, and has exercised a strong influence for the advancement of medical science and education.

To this class of literature this country has contributed a noteworthy share. Excluding those devoted to dentistry, pharmacy, popular hygiene, and "isms" of various kinds, we find that one hundred and ninety-five medical journals have been commenced in this country, including reprints of foreign journals, making in all one thousand six hundred and thirty-seven volumes, or a greater bulk than the text-books and monographs.

Prior to the establishment of medical periodicals, there was little or no encouragement or opportunity for a physician to record his observations. The professor in a medical school might, in an introductory notice to the thesis of one of the students—the so-called programma or propempticon inaugurale—make a statement, not to exceed sixteen pages upon any subject, whether connected with that treated of in the thesis or not, and sometimes such a paper was continued through the programmata of twenty or thirty dif-

---

[1] The figures of this distribution among publishers are only an approximation, and are probably too small, since the publishers' names are not stated in many of the lists of books from which titles have been derived.

ferent dissertations, making it very difficult at the present day to secure the entire work.

But if the country doctor had a communication to make to his brethren, he must either do it by a pamphlet printed at his own expense, or must forward it to some one connected with a medical school or scientific association, and trust to him that it should be made known and recorded. The professors themselves, as was natural, gave the greater part of their thought and labour to their systems, theories, and commentaries.

It was the day of large books, and unless one could produce a volume, he received little encouragement to write. At the present day, the demand for brief papers and reports of single cases, exceeds the supply.

The weekly and monthly periodicals are omnivorous and insatiable in their requests for contributions Through the medical journals have been given to the world nearly all the discoveries which the science and art of medicine owes to American physicians. They furnish the original data which are the foundations of monographs and text-books, and their files remain interesting and valuable when the latter have become obsolete and are forgotten.

Medical journalism in the United States presents some peculiarities, although not nearly so many as is commonly supposed, and has been the subject of severe, and, to some extent, merited criticism; but while it includes some of the worst, it also contains the best of our medical literature, and some details as to its rise, progress, and character, may therefore be of interest.

The first medical journal printed in this country was a selection and translation from the "Journal de Médecine Militaire," issued in Paris from 1782 to 1788. This translation was published in New York about 1790, forming a volume of one hundred and twenty pages 8vo., which is quite rare.[1] The original journal from which this is made up is one that is valuable to the army surgeon; and the reprint is here referred to as being the first medical journal printed in the United States, and because the fact of its existence is probably known to very few.

The first American medical journal was a quarterly, "The Medical Repository," edited by S. L. Mitchell, Edward Miller, and E. H. Smith, and published at New York, from 1797 to 1824. That this met an existing want is shown by the fact that the demand for the earlier volumes was sufficient to warrant the issue of a second edition of the first and second volumes in 1800, and a third edition of the same volumes in 1804–5.

Dr. Elihu H. Smith, the projector of this journal, was born in Connecticut in 1771, graduated at Yale in 1786, and died in 1798. Although so young, he had edited several works, and contributed largely to literary periodicals, as well as to his own medical journal.

Dr. Samuel L. Mitchell, 1764–1831, studied under Dr. Bard, and graduated in medicine at Edinburgh, in 1786. As Professor of Chemistry and Natural History in Columbia College, and from 1820 to 1826 of Materia Medica and Botany, chief editor of the "Medical Repository," representative in Congress in 1801–4, and 1810–13, and United States Senator, 1804–9, he lectured and wrote upon almost all subjects, and his papers are scattered through various periodicals at home and abroad. He was rather a naturalist than a physician, and has very properly been called a "Chaos of Knowledge."

---

[1] "A Journal of the Practice of Medicine, and Surgery and Pharmacy in the Military Hospitals of France. Published by order of the King. Reviewed and digested by M. De Horne, under the inspection of the Royal Society. Annotated from the French by Joseph Brown. No. I., vol. i., New York: J. McLean & Co."

Dr. Edward Miller, 1760–1812, was a native of Delaware, and a graduate of the Medical Department of the University of Pennsylvania in 1789. In 1807 he accepted the chair of the Practice of Physic in the College of Physicians and Surgeons, and in 1809 was appointed one of the Physicians to the New York Hospital. His writings were collected and published in one volume in 1814, the most important being his papers on Yellow Fever.

The idea of the publication of the "Medical Repository" was probably taken from the "Annals of Medicine" of Duncan, a continuation of the "Medical and Philosophical Commentaries of Edinburgh," and of which the "Edinburgh Medical Journal" of the present day is the successor. Although, owing to the tastes of Dr. Mitchell, it contains many dissertations which are now obsolete, the entire set of twenty-three volumes is even to-day well worthy of a place in the physician's library. At its close its subscribers passed to the "New York Medical and Physical Journal," and from that time, New York city has never been without a medical periodical.

Thirty-one medical journals have been commenced in that city, besides nine devoted to specialties, and six reprints of foreign journals. The most important of these, in addition to those already named, are the "American Medical and Philosophical Register," edited by Drs. Hosack and Francis, 1810–14; the "New York Medical Magazine," edited by Mott and Onderdonk, the "New York Journal of Medicine and Surgery," 1839–41, one of the best journals in this country, edited by Drs. Watson and Swett, the "New York Journal of Medicine," edited by Forry, Lee, Stephen Smith, and others, continued as the "American Medical Times," of which the "Medical Record" of to-day may be considered as the representative; the "New York Medical Journal," edited successively by Drs. Hammond, Dunster, and Hunter, 1865–76, and the "Archives of Scientific and Practical

Medicine," edited by Brown-Séquard, 1873, which unfortunately ceased with its fifth number. The "Buffalo Medical Journal," edited by Dr. Austin Flint, 1845-60, and then merged in the "American Medical Monthly," is also a valuable series.

The second medical journal published in this country was the "Philadelphia Medical Museum," edited by Dr. Coxe, 1804-1811, followed almost immediately by the "Philadelphia Medical and Physical Journal," edited by B. S. Barton, and published at irregular intervals, 1804-1809. This journal, as was to be expected from the tastes of its editor, contains a large proportion of articles on natural history. Other well-known journals published in Philadelphia are the "American Medical Recorder," a quarterly, 1818-29, whose subscription list passed to the "American Journal of the Medical Sciences;" the "North American Medical and Surgical Journal," 1826-31; the "Medical Examiner," 1838-56, which united with the "Louisville Review," forming the "North American Medico-Chirurgical Review," 1857-61; the "Medical and Surgical Reporter," 1856-76; the "Photographic Review of Medicine and Surgery," 1870-72; and the "Philadelphia Medical Times," 1870-76.

The most important journal on our list is the "American Journal of the Medical Sciences." This began as the "Philadelphia Journal of the Medical and Physical Sciences," in 1820, under the editorship of Dr. N. Chapman, who is said to have undertaken it under the stimulus of the phrase of Sidney Smith, so often quoted during the past year: "Who reads an American book?" In 1825 a new series began, edited by N. Chapman, W. P. Dewees, and J. D. Godman. This continued until 1827, when Dr. Isaac Hays, who had been associate editor in the last volume—number five of the new, or fourteen of the whole series—took charge of the Journal and gave it its present name. The ninety-seven volumes of this Journal need no eulogy. . They contain many original papers of the highest value; nearly all the real criticisms and reviews which we possess; and such carefully prepared summaries of the progress of medical science, and abstracts and notices of foreign works, that from this file alone, were all other productions of the press for the last fifty years destroyed, it would be possible to reproduce the great majority of the real contributions of the world to medical science during that period. It is evident that its editor has exercised a careful supervision over every part, but his personality is nowhere apparent, there being no editorial articles, and very few papers appearing over his signature.

Baltimore produced the third of our medical journals, the "Baltimore Medical and Physical Recorder," edited by Dr. Tobias Watkins, 1808-9. This only reached number one (1) of the second volume, and it is somewhat curious that of the ten medical journals and one reprint which have been commenced in that city, the duration of each has been comparatively brief. One little known may be referred to, "The Baltimore Philosophical Journal and Review," edited by Dr. J. B. Davidge, of which one number was published in 1823. It contains "a memoir on fractures of the thighbone," and "a case of extirpation of the parotid," each by the editor.

The first medical periodical published in Boston was of a popular character, "The Medical and Agricultural Register," 1806-7. The "New England Journal of Medicine and Surgery" began as a quarterly in 1812, and in 1828 was consolidated with the "Boston Medical Intelligencer," and became a weekly, forming "The Boston Medical and Surgical Journal," which has continued to the present time. The original quarterly was well edited, and contains some valuable papers. Under the editorship of Dr. J. V. C. Smith, which lasted for over fifty volumes, it would seem that no articles were ever refused admission to the weekly.

As stated by Dr. Hunt,[1] "John C. Warren and X. Chabert were received with equal courtesy. In its department of reviews it was most complacent. From Rokitansky to Mrs. Joel Shew all were erudite. On its editorial pages nothing was attacked, everything was conciliated. Legitimate medicine was right to be sure, but the community would appreciate it better if it were not quite so right. Contributors of merit dropped off, and the journal became the receptacle of more 'remarkable cases' than any other was ever blessed with." From the date of this criticism there has been great improvement, and it is to-day one of the best.

The first medical journal west of the Alleghanies was the "Western Quarterly Reporter of Medical, Surgical, and Natural Science," edited by John D. Godman, Cincinnati, 1822–23, which reached number two of the second volume. This was followed by the "Ohio Medical Repository," edited by Guy W. Wright, issued semi-monthly, Cincinnati 1826–27. This has become one of the rarest of American medical journals. The only articles of interest which it contains are a series of papers by Dr. John Locke, on the Medical Botany of the West, and a few reports of cases and contributions to pathological anatomy, by Dr. John P. Harrison. (This journal must not be confounded with another of the same name, published at the same place, in 1835–36.) It was merged into the "Western Medical and Physical Journal," edited by Drs. Daniel Drake and Wright. At the end of the first volume, in 1828, the editors agreed to disagree, and Dr. Wright published one number of a second volume, but the real continuation was issued by Dr. Drake, under the title of the "Western Journal of the Medical and Physical Sciences." This contained some of Dr. Drake's best and most characteristic writings, and forms a valuable and interesting series.

[1] Buffalo Medical Journal, 1856, xii. p. 312.

Two attempts were made by Dr. Eberle to establish a journal at Cincinnati; the first, the "Western Medical Gazette," after one or two suspensions, ceased with the second volume, in 1835; the second, the "Western Quarterly Journal of Practical Medicine," 1837, did not get beyond the first number. "The Western Lancet," edited by L. M. Lawson, continued from 1842 to 1857, when it took the name of "The Cincinnati Lancet and Observer," which is still flourishing. Several medical journals were started at Columbus, only one of which, "The Ohio Medical and Surgical Journal," 1848-64, was successful. A rare medical periodical and curiosity in its way is "The Belmont Medical Journal," published at Bridgeport, Ohio, under the auspices of the Belmont County Medical Society, 1858-60. With this belong the transactions of the same society from 1847 to 1857, forming in all, three small volumes in 12mo. These publications are unique in their way, and illustrate what can be done by a county medical society, composed entirely of country practitioners. They contain some amusing flights of rhetoric, and some well-recorded cases, and many of the papers are interesting because it is evident that they were written precisely as the authors talked.

The first medical journal of Kentucky was the "Transylvania Journal of Medicine," a quarterly, published at Lexington, from 1828 to 1839, forming a series of twelve volumes, of which complete sets are rare and valuable. In 1840 commenced "The Western Journal of Medicine and Surgery," Louisville, 1840-55, which may be considered as a continuation of Dr. Drake's "Western Journal," above referred to, combined with the "Louisville Journal of Medicine and Surgery," edited by Drs. Yandell, Miller, and Bell, in 1838, and of which but two numbers were published.

"The Richmond and Louisville Medical Journal," now in course of publication, edited by Dr. E. S. Gaillard, 1868-

76, is a continuation of the "Richmond Medical Journal," published at Richmond, Va., 1866–68. "The American Practitioner," edited by Drs. D. W. Yandell and T. Parvin, 1870–76, is a continuation of the "Cincinnati Journal of Medicine," commenced in Cincinnati in 1867.

"The Illinois Medical and Surgical Journal" commenced at Chicago in 1844, and has continued to the present time under various names, being now known as "The Chicago Medical Journal and Examiner."

The first journal published west of the Mississippi was "The St. Louis Medical and Surgical Journal," founded by Dr. M. L. Linton, in 1843, which is still in existence.

In the South the first medical periodical was the "Journal de la Société Médicale de la Nouvelle Orleans," a quarterly, published in 1831. A monthly journal of the same name appeared in 1859–61. The most important is the "New Orleans Medical and Surgical Journal," which, with two suspensions, has continued from 1844 to the present time. "The Southern Medical and Surgical Journal," edited by Anthony Eve and others, published at Augusta, forms a series of twenty-one volumes, which contain many valuable cases, papers, and reports. "The Charleston Medical Journal and Review," 1846–60, and 1873–76, is the principal medical periodical of South Carolina.

In Tennessee, "The Nashville Journal of Medicine and Surgery," 1851–61, and 1866–76, and "The Southern Journal of the Medical and Physical Sciences," 1853–57, are worthy of note.

The principal medical journal in Virginia was "The Virginia Medical and Surgical Journal," edited by G. A. Otis and others, Richmond, 1853–61. In the same city was published, during the war, "The Confederate States Medical and Surgical Journal," 1864–65, a quarto sheet containing much valuable data in military surgery. Complete files of this are very rare.

On the Pacific coast eight medical journals, in all, have been commenced, two of which did not get beyond the first number. The oldest one now in existence is "The Pacific Medical and Surgical Journal," which began in 1858.

Five medical journals have been commenced in Michigan, two of which are now in existence.

Connecticut, Iowa, Maine, Minnesota, New Hampshire, New Jersey, Oregon, Vermont, and West Virginia have each had one journal, all of which are now extinct except "The West Virginia Medical Student." Perhaps two may be claimed from Maine, counting "The Journal of the Medical Society of Maine," one number of which was issued at Hallowell in 1834.

Of journals devoted to dentistry there have been about twenty, making one hundred and thirty volumes in all.

The earliest one was the "American Journal of Dental Science," which commenced in New York, in 1839, was suspended from 1860 to 1867, and is still in existence.

In 1876 there are four dental journals in existence in this country, while England has but one, France two, and Germany one.

Of journals devoted to pharmacy, there have been six worth mentioning; the oldest being the present "American Journal of Pharmacy," which began in 1825, as the "Journal of the Philadelphia College of Pharmacy." This journal is by far the most valuable of this class in this country, and is furthermore noteworthy, and to be specially commended for having done what no medical journal in this country has accomplished, namely, the publishing of a complete index for its series, which was done in 1873, and which doubles the practical value of the set. The total number of volumes published of this class is ninety-four.

Besides the regular encyclopedic medical journals, there have been about as many more devoted to "isms" and "pathies," and to popular and family medicine and hygiene, many of these last being merely advertisements.

With the recent development of specialties in medicine, several journals devoted to particular subjects have appeared, and an increase in the number of these may be expected.

In this connection may be mentioned, as a curiosity in literature, a periodical publication devoted to the abuse of an individual physician, namely, the "Rush Light," published in New York in 1800, by William Cobbett, under the pseudonym of Peter Porcupine, for the vilification of Dr. Benjamin Rush. Seven numbers were issued, of which only the first two bore the imprint of place of publication, the last two were printed in London, and a complete set is very rare.

A most powerful agent for the diffusion in this country of the knowledge of the labours and writings of European physicians, has been the republication of the principal English Quarterly Reviews, of "Braithwaite's Retrospect," and of "Ranking's Abstract." To this should be added, perhaps, the so-called "American Edition of the London Lancet," which is a selection rather than a reprint, and the subscription list of which was at one time very large.

Of journals printed in foreign languages, there have been commenced, three in German, three French, and one Spanish. The French journals were all issued at New Orleans: two of the German journals appeared in the State of New York, and one in Philadelphia.

The Spanish journal was intended mainly for circulation in Cuba.[1] Its issue ceased with the third number.

Our medical journals vary so much in character, style, and purpose, that it is hardly possible to make any assertion

---

[1] "Revista Medico-Quirurgica y Dentistica." Quarterly. New York and Havana, 1868.

with regard to the mass which shall be at the same time broad and true. They may be divided into three classes: first, those not connected with any medical school, and which draw their contributions from a wide field, including such as the "American Journal," "The New York Journal," "The Medical Record," "The Medical Times," and "The Boston Medical and Surgical Journal;" second, those which rely for contributions and material mainly on the professors of a medical school and the hospital clinics connected with it, but which are not specially devoted to its interests; third, those which are mainly devoted to advocating the interests of a school, and the attacking rival institutions, and which are, to use Carlyle's phrase, "Windmills put out to catch or take advantage of the wind of popular favour." These journals sometimes contain valuable reports of cases obtained from the college clinics, but the personal editorial element in them is usually in excess, and they are of interest to but a small local circle. To them applies the untranslatable French criticism, "Il y a trop de tintamarre la dedans, trop de brouillamini."

Of the first class, some compare favourably with the best of the journals of other countries: of the last class, some are as bad as, but not worse than, the worst. Comparatively few persons are acquainted with the poorer class of foreign medical journals, published in the smaller towns of the provinces, which have most of the defects which are so strongly condemned in some of our own publications as if they were unique.

The reports to the American Medical Association, by its committees on American Medical Literature, devote much space to periodicals, and contain many judicious criticisms upon their defects and errors. A common complaint is that there are too many. The reply to this is usually that of Dr. Drake, that it is desirable that the country practitioners be induced to write, and that one means of doing this is the diffused localization of journals. This is due to the fact that inexperienced and modest men will furnish an article or report to a journal in their immediate neighbourhood, with whose editor they are personally acquainted, while they would not do so to one at a distance.

The number of subscribers to the greater number of our journals is small, the issue being, for many, less than a thousand, and, for some, hardly five hundred copies.

The motive for the existence of the minor journals is not for direct profit, but as an indirect advertisement for certain individuals, or—and this is more common—the desire to have a place in which the editor can speak his mind and attack his adversaries without restraint. The defects in the medical journals are, to a certain extent, the characteristic ones in our medical literature, and are chargeable mainly to the lack of general education and mental culture in the majority of readers whose tastes are to be accommodated. An urgent want of many of the subscribers is a sort of continuation of the course of education given in the schools. We find, for instance, in the pages of some medical journals, articles which make no pretensions to originality, but are simply didactic lectures to a class *in absentia*. The defects in the so-called original contributions are, for the most part, due to imperfect education in the writers, and betray, not merely an ignorance of facts previously ascertained and recorded, but defective mental training and an inability to comprehend the relations of the facts which are known, the result of which is the stringing out of a series of irrelevant and tedious details, and, in the attempts at deduction, the production either of vague and valueless generalizations, or conclusions which do not follow from the premises. As an illustration, take the majority of the articles which have appeared on a disease which would seem to be peculiar to this country, viz. the so-called "milk-sickness" or "trembles."

Since the first notice on this affection in Dr. Drake's

Notices of Cincinnati, in 1809, there have been printed four pamphlets and one hundred and ten (110) articles in journals and transactions, on this subject. Yet it cannot be said to-day, that we have any definite knowledge as to the pathology or causes of this affection, or that, so far as man is concerned, we are absolutely certain that there is any special disease which should be thus named, as being caused by the milk, or flesh of cattle affected with the "trembles." It has been said to be caused by certain plants, yet no scientific experiments have been made on the effects of these plants. No attempt has been made to produce the disease in an animal remote from infested localities, by the use of the suspected plants, or better, by the use of an extract containing their active principles; no chemical or microscopical examinations have been made, in short, we have nothing but an account of symptoms, and much of that is from hearsay.

Many articles intended to be practical, are very far from being such, although the authors would probably be surprised and indignant to hear them termed otherwise. They profess to give the results of the writer's personal experience with a certain disease, but this disease is only named, not described, and the gross results only are given, that is to say, we are told how many recovered. The object of such writers, to use their own words, is to tell us "what is good for biliousness, or low fever, or pneumonia." Their productions read curiously, like the literature of the last century, and are to be classed with old women's advice; amusing generally; practically suggestive sometimes; clear, scientific, and conclusive, never.

The so-called clinical lectures, and reports of cases and operations, are of two kinds. When properly prepared they are most useful and valuable, and are the best contributions to a journal which the majority of physicians can make, although by no means the highest class of medical literature. But a large number of such articles as are published, are simply padding, worse than useless, since their titles become a part of the bibliography of medicine, compelling each succeeding inquirer to refer to them, or risk the loss of some really valuable reference.

We have reached that stage of development, when it is in no way desirable that we should be informed that one dislocated shoulder was reduced, one leg amputated, and two hare-lips operated upon, not even if the usual text-book explanations are added, so as to make up the five or six pages of the report of a college clinic. We have had enough reports of specimens of "Aneurism of the Aorta," or "Medullary Sarcoma," or "Tumour of the Breast," in which little or no information is given with regard to the symptoms during life, and the principal fact stated is the size or weight of the specimen.

It is a useless case of labour which lingers through three or four pages, to terminate in the usual manner with the stale old moral about "meddlesome midwifery," and it is at once amusing, exasperating and pathetic, to glance over the "contributions from the clinic" of the young specialist who has set to work to write himself into notice, not in a journal devoted to his specialty, but in one of the encyclopedic periodicals, having been instructed that this is "legitimate advertising."

"Medical journalism is not a profession in this country. With one or two exceptions, our medical editors are engaged in practice and lecturing, and their labour in connection with the journals is not directly remunerative, nor is it the main object of their thoughts." The result of this appears in that large section of almost every journal which is devoted to reviews, abstracts, news items, etc. Nevertheless, as we have before stated, our medical journals are the most important and valuable part of our medical literature, and it is mainly in and by them that improvement may be hoped for and effected.

At the beginning of 1876, there were in course of publication throughout the world about 280 regular medical journals. Of this number, Germany and Austria had 57; France 52; Great Britain, not including her Colonies 29; the United States 46; Italy 31; Belgium 8; Mexico 8; Canada 7; Holland 6; Spain 6. As to the form of publication, the United States has the largest proportion of monthlies, and France and Germany of weeklies and bi-weeklies.

The proportion of periodical to other forms of medical literature is in excess in this country, as will be clearly seen if we compare the number of medical books published in the several countries. Taking the "Bibliotheca Medico-Chirurgica," of Ruprecht, for the years 1874–75, and counting the publications noted in it, excluding journals, pamphlets, and popular and irregular works, we find that the United States is credited with 55 volumes; England 179; France 409; Germany 419; Italy 120; Spain and Portugal 104. If we count only first editions of original works, we find that the United States has published during these two years 36; England 92; France 314; Germany 288; Italy 88; and Spain and Portugal 30.

These figures are, of course, not exact, but the proportions shown are probably nearly correct. Taking the number of volumes of medical publications of all nations, excluding journals, for these two years, the United States has published about six per cent. of the whole, certainly not the quantity which should have been produced if everything was as it should be.

Medical Societies.—An important influence upon the progress of medicine, and the relations of physicians to each other, and to the public, has been exerted by our medical societies, some of which date from the last century, and which are found almost everywhere. The first State medical societies, such as those of New Jersey, Massachusetts, Delaware, New York, etc., were charged with the duty of licensing persons to practise medicine, to which license an examination was a necessary preliminary. In this way these societies were the principal agents in fixing the standard of medical education, and although after the establishment of medical schools the diploma of one in good repute was accepted in lieu of an examination, this was by courtesy rather than law, and made it necessary that the standard of the schools should at least be equal to that prescribed by the society. For convenience of reference, we give a list of the most important medical societies of the United States, arranged in alphabetical order by States:—

	Organized.	First publication.	No. of vols. of publications.
American Medical Association	1847	1848	27
American Ophthalmological Society	1864	1865	7
American Otological Society	1868	1869	1
American Pharmaceutical Association	1852	1852	24
American Public Health Association	1872	1875	1
National Quarantine and Sanitary Convention	1857	1857	4
Medical Association of the State of Alabama	1847	1848	19
State Medical Association of Arkansas	1870	1871	5
Medical Society of the State of California	1870	1870	5
Territorial Medical Society of Colorado	1871	1872	5
Connecticut State Medical Society	1792	1844	20
Medical Society of Delaware	1789	......	...
Medical Society of the District of Columbia	1833	1874	2
Clinico-Pathological Society of Washington	1865	......	...
Florida Medical Association	1874	1875	1
Georgia Medical Association	1849	1850	20
Georgia Medical Society of Savannah	1804	......	...
Illinois State Medical Society	1851	1851	23
Drake Academy of Medicine	1872	1874	1
Indiana State Medical Society	1849	1849	27
Iowa State Medical Society	1850	1850	10
Medical Society of the State of Kansas	1858	1867	2
McDowell Medical Society	1874	1875	1
Kentucky State Medical Society	1851	1851	19
Société Médicale de la Nouvelle Orleans	1812	1831	3
Medical Society of the State of Maine	1834	1834	1
Maine Medical Association	1853	1853	6

LITERATURE AND INSTITUTIONS. 345

	Organized.	First publication.	No. of vols of publications.
Medical and Chirurgical Faculty of Maryland	1789	1853	4
Boston Society for Medical Improvement	1828	1853	5
Boston Society for Medical Observation	1846	......	...
Boylston Medical Society	1811	70 prize essays published in journals.	
Gynæcological Society of Boston	1869	1869	5
Massachusetts Medical Society	1781	1790	41
Michigan State Medical Society	1819	1850	15
Minnesota State Medical Society	1855	1870	6
Medical Association of the State of Mississippi	1856	1870	1
Medical Society of the State of Missouri	1850	1850	12
Nebraska State Medical Society	1868	1869	6
New Hampshire Medical Society	1791	1854	21
New Jersey State Medical Society	1766	1859	17
Medical Association of Southern Central New York	1847	1848	11
Medical Society of the County of Albany	1806	1864	2
Medical Society of the County of Kings	1822	1858	2
Medical Society of the County of New York	1806	......	...
Medical Society of the State of New York	1807	1808	34
Medico-Legal Society of New York	1867	1874	1
New York Academy of Medicine	1847	1851	8
New York Medical Journal Association	1864	......	...
Pathological Society of New York	1844	......	...
Physico-Medical Society of New York	1815	1817	1
Medical Society of the State of North Carolina	1850	1850	22
Academy of Medicine of Cincinnati	1857	......	...
General Medical Society of Ohio	1827	1829	2
Medical Convention of Ohio	1835	1835	13
Ohio State Medical Society	1846	1850	26
Belmont Medical Society	1847	1848	8
Medical Society of the State of Oregon	1874	......	...
College of Physicians of Philadelphia	1787	1793	11
Medical Society of the State of Pennsylvania	1848	1851	18
Pathological Society of Philadelphia	1857	1869	4
Philadelphia County Medical Society	1849	......	...
Philadelphia Obstetrical Society	1868	1873	3
Rhode Island Medical Society	1812	1859	1
Medical Society of South Carolina	1789	......	...
South Carolina Medical Association	1848	1849	16
Tennessee State Medical Society	1830	......	...
Medical Association State of Texas	1869	1869	4

	Organ-ized.	First publica-tion.	No. of vols. of publica-tions.
Medical Society of the State of Vermont	1814	1864	4
Medical Society of Virginia	1821	1871	5
Medical Society of Washington Territory	1873	1873	3
Medical Society of the State of West Virginia	1867	1868	8
Wisconsin State Medical Society	1842	1856	9

The formation of the American Medical Association was due to a wide spread and loudly expressed dissatisfaction on the part of the leading physicians of the country, with the low standard of medical education, and to a general conviction that the remedy for this lay neither with the schools nor the State medical societies. It was hoped that by forming an association which should represent all parties interested, a sufficient pressure of opinion might be brought to bear upon physicians and upon the schools, to secure the return to the requirements for graduation of the earlier medical colleges. After one or two futile attempts, the New York State Medical Society set on foot a movement which resulted in a meeting of a convention in the city of New York, in the year 1847, in which were present representatives of medical societies and colleges from sixteen States. A similar convention met the following year in Philadelphia, at which the title, by which it is now known, was assumed. The series of its annual volumes of transactions contains some reports and papers of much value and interest, mingled with much that is unworthy of publication under the auspices of our National Medical Society, or indeed of any other. Many of the reports of the chairmen of the several committees are of permanent historical value. Its most valuable contribution to our literature, has been the publication of a code of ethics, which is, theoretically at least, accepted as authoritative throughout the United States, and which, although some of its provisions have been objected to, is, as a whole, the most satisfactory exposition in existence of the proper relations of physicians to each other, and to the public.

Of late years, the original purpose of this association has been to some extent departed from.

It was not primarily intended to promote literature or scientific research, or to afford a means of publication for writers. Our national and State medical societies have been mainly useful as social gatherings, promoting acquaintance, and the feeling of professional brotherhood and *ésprit de corps* among their numbers, and as giving the means for agreement, and the expression of opinion, upon questions relating to education, ethics, etc.; by that large body of physicians engaged in general practice, who do not write or lecture, but simply vote. As sources of addition to the science and literature of medicine, they do not play a conspicuous part, nor is it easy to see how it can be otherwise; the real discovery, the carefully prepared paper, the description of a new symptom, pathological appearance or remedy are not usually communicated to such societies. No effectual supervision as to quality of papers which may be read or printed can, or at least will be exercised by committees, and a communication which a first class medical journal has "declined with thanks," may be taken to the State, and even to the National Society with a reasonable certainty that it can be made to appear in the transactions. The discussions on papers in such associations seldom have any scientific value, from want of special preparation on the part of the speakers, although they are sometimes amusing, and, to use an expressive word, "spicy," from the use of personalities. Whether this state of things can be improved is doubtful, though attempts to do so are of course commendable.[1]

---

[1] The best suggestion to this end for the American Medical Association which I have heard is that each section should elect its own officers and members, and should be managed by a special

The journals have to a great extent superseded the necessity of using societies as a means of publication, and the best work of such associations seems to consist in bringing the leaders into personal relation with the mass of the profession, and in serving as courts of arbitration and appeal, where local difficulties can be adjusted, and whose decisions will command the assent of the majority of their members.

The Transactions of the New York State Medical Society were, for a number of years, published by the State, which proved, upon the whole, to be not a desirable mode of issue, and the last volume, published by the society itself, is a great improvement upon its predecessors. What such societies might do is shown by the paper of Dr. Thomas C. Brinsmade, giving an accurate record of his practice for twenty-one years. This makes 300 pages of the volume of the Transactions of the New York State Medical Society for 1858, and contains carefully analyzed statistics of 37,872 cases. This had been preceded in 1851 by an elaborate account of the medical topography of the city of Troy, his place of residence. Taken together, these papers are very valuable, and set an example of a mode of adding to the store of medical knowledge, which is within the power of every practitioner.

An interesting experiment is now in course of trial in Alabama, where the State Medical Society has been made

committee who shall designate the subject for discussion, and the leaders in debate. If the members of the committee each year are selected from a single city, it would have an additional advantage. For instance, let the managing committee of the surgical section be, this year, all residents in New York city, while Boston takes obstetrics, Philadelphia practice, etc. The next year New York can take practice, Chicago surgery, etc. In other words, transfer all the responsibility for scientific work to the sections, and let these sections be organized and managed systematically to that end alone.

the State Board of Health, and the official adviser of the Legislature in all matters pertaining to public hygiene.

The American Public Health Association, organized in 1872, may now be considered as fairly established. The operations of this society have special interest to the medical profession, since it may become an important means of educating the public, and enabling it to distinguish between the scientific physician and the ignorant pretender.

We have another class of medical societies which require an abundance of clinical and pathological material; members actually engaged in original investigations, and frequent meetings, as conditions for usefulness and success. As a rule, these can only exist in large cities, where they exert a powerful influence and stimulus to exertion on their individual members. It must be admitted that our societies of this kind seldom bring out the best work of their members, and that such discussions as occur in similar societies in London and Paris, continued week after week, and even month after month, for which elaborate preparation is made by the speakers, and in which the results of clinical observation and extensive literary research are rendered attractive and striking, by splendor of diction and perfection of style, are very rare.

The most important of these societies are the College of Physicians of Philadelphia, the New York Academy of Medicine, the pathological societies of Philadelphia and New York, the Boston Society for Medical Improvement, and certain societies devoted mainly to specialties. Among these should be mentioned the Medico-Legal Society of New York, organized in 1867. In 1874 it published a volume of papers relating to medical jurisprudence, which will be followed by others. It is also forming a valuable library in its own department, and has been the means of bringing the members of the medical and legal professions of New York to better acquaintance with each other. It is but justice to say that much of its good work and prosperity is due to the energy of its late president, a prominent lawyer, Mr. Clark Bell.

The majority of our physicians are, and must be, content to leave to a few special workers the labour and pleasure of sifting and selecting from the original sources of medical literature, having neither the wish nor the power to examine for themselves the works of the great leaders and teachers of times past, or the mass of books and pamphlets which are daily streaming from the press; but there is nevertheless among them a fair amount of appreciation of the value and necessity of such work, and of the usefulness and desirability of collections of the records of their science. During the last ten years, the writer has had occasion to examine many private libraries of physicians in all parts of the country, in country villages as well as the large cities, and it has been a matter of surprise and pleasure to find so much interest taken in subjects relating to the history and bibliography of medicine by men remote from large libraries, and without the stimulus of companionship in, and sympathy with such tastes. And it will usually be found that the physician who has on his shelves half a dozen old folios and quartos, including perhaps copies of Sydenham, Morgagni, and Van Swieten, is a man of more culture and broader views than the one who has only the modern manuals, or rather those which were modern when he attended lectures.

Until recently few of our writers have made much use of bibliographical research. We now have public medical libraries in this country, which afford to the student and scholar good facilities for research, and which bid fair, at no distant day, to rival in magnitude and practical working value, if not in manuscripts and incunabulæ, the best in the old world.

Philadelphia has several libraries of much interest and value to the medical bibliographer and scholar. The oldest medical library in this country is that of the Pennsylvania Hospital, founded in 1762, and now containing about 13,000 volumes, many of which were selected for the hospital by Doctors Lettsom, in London, and Louis, in Paris. Its classed catalogue, issued in 1857, is a valuable work of reference.

The library of the College of Physicians, of Philadelphia, which dates from 1788, now contains about 19,000 volumes well selected, receives about eighty current journals, and, next to the library at Washington, is the most valuable collection of the kind in this country. Much of its prosperity and excellence is due to Doctor Samuel Lewis, whose donations, amounting to several thousand volumes of choice books, are kept in a room by themselves, and known as the "Lewis Library." The great want of this library is a good printed catalogue, which would double its value and usefulness. The medical part of the Loganian Section in the Philadelphia Library contains about 1800 volumes, mostly old and rare. These three libraries supplement each other to a great extent, there being probably not less than 26,000 volumes between them, which are not duplicates. The fourth library is at the University of Pennsylvania, in West Philadelphia, and contains about 3000 volumes, the gift of Dr. Alfred Stillé. It may be noted here that almost all attempts to establish medical libraries in connection with medical schools have been failures. Commenced with enthusiasm, they soon become antiquated, are rarely consulted, except by one or two species of beetles, are never properly catalogued or cared for, and dust and mould reign in them supreme. Students and teachers want the newest books and journals only. Libraries are used by the scholar and author, and for such are the true universities.

In New York, the library of the New York Hospital is the largest of its class, containing about 10,000 volumes. An excellent foundation for a library has been acquired by the Academy of Medicine, by the gift from Dr. Purple of a complete file of regular American medical journals and of a large number of rare pamphlets. The collections of journals of the Medical Journal Association of New York, and the German Dispensary are valuable sources of information to the student.

The Boston Public Library has at present the best collection of medical books in that city, numbering about 11,000 volumes, for the most part standard works and periodicals. Its usefulness is much diminished from the want of a good printed catalogue of this section. The library of Harvard College contains between 5000 and 6000 volumes on medicine; and the Treadwell Library, at the Massachusetts General Hospital, has about 5000 volumes. The medical library of most promise in Boston is that of the Medical Library Association, which, though only a year old, has about 3000 volumes, and will probably rapidly increase.

In Cincinnati the City Hospital has a fair collection. The Mussey Medical and Scientific Library, at present, is a special deposit in the Cincinnati public library, and contains about 4000 volumes and 2000 pamphlets.

The National Medical Library at Washington, under the direction of the Surgeon-General of the Army, contains 40,000 volumes, and about the same number of pamphlets. It has been formed within the last twelve years, and the use that is made of it by physicians from all parts of the country, and the general and strong interest that is felt in its progress affords satisfactory evidence, if such were needed, that it meets a want of the profession. Its subject catalogue is nearly ready for the press.

Besides these public libraries, there are several valuable private collections of medical works in this country, some of which have been already given to public use, such as those of Drs. Purple, Stillé, and Mussey, already referred to. Two others are worthy of special mention, the first being that of Dr. G. J. Fisher, of Sing Sing, which is rich in the classics of medicine; and the second, that of Dr. J. M. Toner, of Washington, which is especially devoted to American medical literature, and contains many rare pamphlets, besides a nearly complete file of American medical journals. In connection with this last, there is nearly ready for the press a complete index. Besides these, there are a number of valuable private medical libraries in this country, ranging from 1000 to 8000 volumes, and the number of foreign works imported, and the taste for original editions is steadily increasing. It is now possible to verify in this country the majority of the references made by European medical authors, and it is no longer necessary to make costly importations, or to visit Europe to obtain literary data.

With the libraries should be classed the *medical museums*, of which several of much interest and importance have been formed in the United States, for the most part in connection with medical societies and hospitals. The catalogues of these collections, when properly prepared, are very useful books of reference, and some excellent work of this kind has been accomplished, such as the Catalogues of the Warren Anatomical Museum of Harvard, and of the Museum of the Boston Society for Medical Improvement, each by Dr. J. B. S. Jackson; of the Pathological Museum of the Pennsylvania Hospital, by Dr. Wm. Pepper; of the Pathological Cabinet of the New York Hospital, by Dr. Ray; and of the Army Medical Museum at Washington, by Drs. Woodhull, Curtis, and Woodward.

The College of Physicians of Philadelphia has a valuable collection, including the Mütter Museum, and a series of unique preparations by Hyrtl.

The practical value of large special museums in connection with good libraries devoted to the same specialities is great, but they are useful rather to the educated physician than to the student; and the numerous small collections which are scattered over the country, in hospitals and private cabinets, are simply so much wasted and unused material, in a scientific point of view, and, though gratifying to the owner as trophies or mementoes, are of little more real use than the strings of teeth which the barbers of old hung out as signs of their skill.

The value of a single specimen of any lesion is usually very small; it is only when they can be brought together by scores and compared that useful and reliable results can be hoped for. As we get older and wiser, we shall probably have fewer journals, medical schools, museums, and libraries than we now possess, for all these means of culture, to have the best effect, require concentration.

Although the permanent importance of oral teaching has, to some extent, been diminished by the diffusion of periodical literature, since the latest discovery or theory can now be promptly made known to those remote from the great centres of learning, the increased use made of clinical instruction, and the necessity for practical demonstration of instrumental methods of diagnosis, have in a great degree compensated for this.

The medical history of a country cannot be considered complete without some account of its *medical schools*, but we have space for little more than a list of those which have flourished in the United States.

The following table gives a list of the regular chartered medical schools of this country, which have had the power of conferring the degree of doctor of medicine, with the date of first graduating class, date of cessation, and number of graduates to the spring of 1876, so far as it has been possible to obtain the data:—

It is possible that a few minor schools of short duration have been overlooked, but such must have been of small importance. No note is made in the list of the various changes of name which some of the schools have assumed.

The number of graduates has been obtained by collation of all the catalogues that could be obtained, and by correspondence. From these data an estimate has been made for the missing years, and the limit of error in the total does not probably exceed one-half of one per cent. It should be observed that little reliance can be placed upon many of the catalogues as to the number of students in attendance, and there are some discrepancies even as to graduates.

Name.	Year of first graduation.	Date of cessation.	Total No. of graduates.
Alabama.			
Medical College of Alabama [Mobile]	1860	......	203
California.			
Medical College of the Pacific, Med. Dept. of University (City) College [San Francisco]	1859	......	90
University of California, Med. Dept. of (Toland Hall) [San Francisco]	1865	......	86
Connecticut.			
Yale College, Med. Dept. of [New Haven]	1814	......	899
District of Columbia.			
National Medical College, Med. Dept. of Columbian University [Washington]	1826	......	427
Georgetown University, Med. Dept. of [Washington]	1852	......	387
Howard University, Med. Dept. of [Washington]	1871	......	37
Georgia.			
Medical College of Georgia [Augusta]	1833	......	1278
Savannah Medical College [Savannah]	1854	......	140
Atlanta Medical College [Atlanta]	1855	......	560
Oglethorpe Medical College [Savannah]	1856	1861	86
Illinois.			
Rush Medical College, Med. Dept. of University of Chicago [Chicago]	1844	......	1786
Illinois College, Med. Dept. of [Jacksonville]	1848	1848	39
Rock Island Medical School [Rock Island]	1849	1849	19
Chicago Medical College, Med. Dept. of Northwestern University [Chicago]	1860	......	481

356   LITERATURE AND INSTITUTIONS.

Name.	Year of first graduation.	Date of cessation.	Total No. of graduates.
Indiana.			
Indiana Medical College, Med. Depart. of Laporte University [Laporte]	1842	1851	136
Medical College of Evansville [Evansville]	1850	......	74
Indiana Central Medical College [Indianapolis]	1850	1852	39
Indiana Medical College [Indianapolis]	1870	......	251
Indiana College of Physicians and Surgeons [Indianapolis]	1875	......	......
Iowa.			
College of Physicians and Surgeons [Keokuk]	1850	......	777
Iowa State University, Med. Dept. of [Iowa City]	1871	......	111
Kentucky.			
Transylvania University, Med. Dept. of [Lexington]	1818	1859	1860
University of Louisville, Med. Dept. of [Louisville]	1838	......	2395
Kentucky School of Medicine [Louisville]	1851	......	520
Louisville Medical College [Louisville]	1870	......	402
Hospital College of Medicine, Med. Dept. of Central University [Louisville]	1875	......	91
Louisiana.			
University of Louisiana, Med. Dept. of [New Orleans]	1835	......	1703
New Orleans School of Medicine [New Orleans]	1857	1870	397
Charity Hospital Medical College of New Orleans [New Orleans]	1876	......	10
Maine.			
Bowdoin College and Med. School of Maine [Brunswick]	1821	......	1137
Maryland.			
University of Maryland, Med. Dept. of [Baltimore]	1811	......	3104
Washington University, School of Medicine [Baltimore]	1828	......	680
College of Physicians and Surgeons [Baltimore]	1873	......	118

LITERATURE AND INSTITUTIONS.   357

Name.	Year of first graduation.	Date of cessation.	Total No. of graduates.
Massachusetts.			
Harvard University, Med. Dept. of [Boston]	1785	......	2206
Berkshire Medical College, [Pittsfield]	1823	1867	1136
Michigan.			
University of Michigan, Med. Dept. of [Ann Arbor]	1851	......	1405
Detroit Medical College [Detroit]	1869	......	204
Missouri.			
Missouri Medical College [St. Louis]	1841	......	921
St. Louis Medical College [St. Louis]	1843	......	1293
Humboldt Medical College [St. Louis]	1867	1869	16
Kansas City College of Physicians and Surgeons	1870	......	46
St. Louis College of Physicians and Surgeons [St. Louis]	1870	1870	8
New Hampshire.			
Dartmouth College, Medical School of [Hanover]	1798	......	1283
New York.			
College of Physicians and Surgeons of the City of New York, Med. Dept. of Columbia College [N. Y. City]	1769	......	3179
College of Physicians and Surgeons of the Western District of New York [Fairfield]	1816	1840	585
Geneva College (Rutgers Med. Faculty) [N. Y. City]	1827 ✓	1830	104
Geneva Medical College [Geneva]	1835	1872	849
Albany Medical College [Albany]	1839	......	1287
University of the City of New York, Medical Dept. of [N. Y. City]	1842	......	3393
University of Buffalo, Med. Dept. of [Buffalo]	1847	......	848
New York Medical College and Charity Hospital [N. Y. City]	1851	1864	310
Long Island College Hospital [Brooklyn]	1860	......	531
Bellevue Hospital Medical College [N. Y. City]	1862	......	1908
College of Medicine of Syracuse University [Syracuse]	1873	......	26

358   LITERATURE AND INSTITUTIONS.

Name.	Year of first graduation.	Date of cessation.	Total No. of graduates.
Ohio.			
Medical College of Ohio [Cincinnati]	1821	......	2170
Cincinnati College, Med. Dept. of [Cincinnati]	1836	1839	95
Starling Medical College [Columbus]	1836	......	887
Cleveland Medical College, Med. Dept. of Western Reserve College at Hudson [Cleveland]	1844	......	1162
Cincinnati College of Med. and Surgery [Cincinnati]	1852	......	760
Miami Medical College [Cincinnati]	1853	......	578
University of Wooster, Med. Dept. of [Cleveland]	1865	......	328
Oregon.			
Willamette University, Med. Dept. of [Salem]	1867	......	63
Pennsylvania.			
University of Pennsylvania, Med. Dept. of [Philadelphia]	1768	......	8845
College of Philadelphia [Philadelphia]	1790	1791	10
Jefferson Medical College [Philadelphia]	1826	......	6668
Pennsylvania College at Gettysburg, Med. Dept. of [Philadelphia]	1840	1861	769
Franklin Med. College of Philadelphia [Philadelphia]	1847	1849	25
Philadelphia College of Medicine [Philadelphia]	1847	1859	502
Rhode Island.			
Brown University, Medical School of [Providence]	1814	1826	68
South Carolina.			
Medical School of the State of South-Carolina [Charleston]	1825	......	2439
University of South Carolina, Med. Dept. of [Columbia]	1868	......	26
Tennessee.			
Memphis Medical College [Memphis]	1847	1873	231
University of Nashville, Med. Dept. of [Nashville]	1852	......	1741
Shelby Medical College [Nashville]	1859	1861	30
Vanderbilt University, Med. Dept of [Nashville]	1875	......	75

Name.	Year of first graduation.	Date of cessation.	Total No. of graduates.
Texas.			
Galveston Medical College [Galveston]	1866	......	123
Texas Medical College and Hospital [Galveston]	1874	......	38
Vermont.			
Castleton Medical College [Castleton]	1820	1861	1449
University of Vermont and State Agricultural College, Med. Dept. of [Burlington]	1823	......	573
Vermont Medical College [Woodstock]	1830	1860	575
Virginia.			
University of Virginia, Med. Dept. of [Charlottesville]	1828	......	533
Medical College of Virginia [Richmond]	1839	......	947
Winchester Medical College [Winchester]	1846	1862	75
Total			73,588

If we take the number of graduates by decades of years during the present century, the result is as follows:—

Years.	No. of graduates.	Years.	No. of graduates.
1769–1799	221	1840–1849	11,828
1800–1809	343	1850–1859	17,213
1810–1819	1,375	1860–1869	16,717
1820–1829	4,338	1870–1876	14,704
1830–1839	6,849		

The first medical school in this country was established by Drs. John Morgan and William Shippen at Philadelphia in 1765, and is now known as the Medical Department of the University of Pennsylvania. From its halls have graduated the majority of the distinguished medical writers, teachers, and practitioners of the United States, and the names of its professors have become household words.

Organized upon the plan of the Edinburgh Medical School, of which its founders were graduates, it has been the model and pattern by which all our medical colleges have been shaped. Its largest graduating class was in 1849, numbering 191. In the following year Professor Chapman resigned, and for the next ten years the Jefferson School graduated the greater number, reaching its maximum of 269 in 1854. The Jefferson Medical College was founded in 1824, under the charter of Jefferson College in Canonsburg, Pennsylvania. The first course of lectures was delivered in 1825–26, the Faculty being Drs. Eberle, McClellan, Rhees, Green, and Beattie. Numerous changes were made in professors, and its classes varied much in size until 1841, when all the chairs were vacated and refilled by Drs. Dunglison, J. K. Mitchell, Pancoast, R. M. Huston, Mütter, Meigs, and Bache. This Faculty continued until 1856, when Professor S. D. Gross succeeded Dr. Mütter. In 1857 Dr. T. B. Mitchell took the place of Dr. Huston, and in 1858 Dr. Dickson that of Dr. J. K. Mitchell.

The second medical school founded in this country was at New York, under the charter of King's College, in 1767. This school has had many vicissitudes, but is now in a flourishing condition, and known as the College of Physicians and Surgeons of the City of New York, being the Medical Department of Columbia College. Its largest graduating class was 110 in 1875.

The Medical Department of Harvard University was founded by Dr. John Warren in 1782. Its maximum class of graduates was 99 in 1866. Recently it has led the way in elevating the standard of medical education, by extending its curriculum to three years, establishing a graded course, and by having decided to institute a real examination into the preliminary education of its students. This has of course diminished its classes somewhat, but no one can doubt that the decision to aim at quality instead of quantity is a wise one, and will in the fulness of time receive its due reward.

The first medical school in the West was established in Lexington, Ky. So early as 1799 a Medical Department was added to Transylvania University, Dr. Samuel Brown being appointed the first professor. Various appointments in the Medical Faculty were made, and a few partial courses of lectures were delivered, but the first full course was not given until 1817, and the degree of M.D. was first conferred in 1818. The founders of the school were Drs. Dudley and Caldwell. Its period of greatest prosperity was from 1830 to 1837, at which last date a disruption took place, and a part of the Faculty removed to Louisville.

The Medical Department of the University of Louisville began as the Medical Institute, chartered in 1833. Nothing was accomplished, however, until the quarrel in the Transylvania School above referred to took place, when Dr. Caldwell enlisted in the cause of the Louisville School, and in 1837 succeeded in obtaining for it a grant of a square of ground, and money for buildings and apparatus. Lectures began the same year, the Faculty consisting of Drs. Caldwell, Cooke, and Yandell, from the Lexington School, and of Drs. Cobb, Henry Miller, and J. B. Flint. In 1839 Dr. Drake joined the School, and in the following year Dr. S. D. Gross took the place of Dr. Flint. In 1846 the School was transferred to the University, and in 1874 it had 123 graduates, its largest class.

In connection with these schools a special reference is due to Dr. Charles Caldwell, their principal promoter. He was of Irish descent, born in North Carolina in 1772; died 1853. After obtaining the best education which his native State could afford, he went to Philadelphia in 1792, and continued the study of medicine under Dr. Rush, passing his examination in 1794, and taking his diploma in 1796. During the next twenty years his pen was constantly busy with lectures, addresses, and controversial articles, many of which related to yellow fever. In 1819 he accepted an invitation to the Transylvania School, and from this time he gave his best energies to this institution, and subsequently to the Louisville School. He was one of the most voluminous writers which this country has produced, but he contributed little or nothing of permanent or scientific value to the literature of his profession, and the only work of his which is worth perusal to-day is his autobiography. His critical reviews, being dictated almost exclusively by personal prejudices, are in almost all cases samples of special pleading rather than true criticism, and characterized by their "smartness" rather than their justice.

In the South the Medical College of South Carolina, chartered in 1823, leads the way. The Medical College of Louisiana was incorporated in 1835, and in 1845 became the Medical Department of the University of Louisiana. This school is remarkable as having received State aid to the amount of $121,000.

In connection with the medical schools, notice should be taken of the Medical Institute of Philadelphia, otherwise known as the Summer School, which, in addition to furnishing instruction to students and supplementing the winter course, was of very great value as a training school for Professors. It was founded in 1817 by Dr. Chapman, and with it were connected, from time to time, Drs. Chapman, Horner, Dewees, Samuel Jackson, J. K. Mitchell, John Bell, Hodge, Neill, Gibson, Gerhard, Norris, and Pepper.

The total number of graduates from our medical schools during the five years ending July 1, 1875, was about 10,250, that is, a little over 2000 per year; the number in 1875 being about 400 more than in 1871.

Dr. J. M. Toner estimated the average age of beginning practice to be $24\frac{1}{2}$ years, of death 58 years, making an average of about 34 years practice to each.[1]

Dr. S. E. Chaillé estimates that there are about 47,000 regular physicians in the United States, being about one to every 700 of the population.[2]

---

[1] Statistical Sketch, etc. Indiana Journ. of Med., 1873, vol. iv. p 1.

[2] The Medical Colleges, etc. New Orleans Med. and Surg. Journ., 1874, vol. i. N. S. p. 818.

Space is wanting for further details with regard to our medical schools. That there are too many of them is a general complaint, the answer to which is the same as that given above with regard to the like objection with regard to medical journals, and which answer is of about the same value in each case.

In attempting to estimate the quantity and value of the additions made by the medical profession of this country to the world's stock of knowledge of the laws of healthy and diseased action, and the means of modifying these actions, it is very difficult to make generalizations which shall be at once clear, comprehensive, and correct. This difficulty becomes an impossibility, if we are to speak of the education, mental characteristics, and professional qualifications of the whole body of physicians of this, or any other country, since only the most vague and indefinite statements will hold good. We have had, and still have, a very few men who love science for its own sake, whose chief pleasure is in original investigations, and to whom the practice of their profession is mainly, or only, of interest as furnishing material for observation and comparison. Such men are to be found for the most part only in large cities where libraries, hospitals, and laboratories are available for their needs, although some of them have preferred the smaller towns and villages as fields of labour. The work of our physicians of this class has been for the most part fragmentary, and is found in scattered papers and essays which have been pointed out in preceding essays; but buds and flowers, rather than ripened fruit, are what we have to offer. Of the highest grade of this class we have thus far produced no specimens; the John Hunter, or Virchow, of the United States, has not yet given any sign of existence.

We have in our cities, great and small, a much larger class of physicians whose principal object is to obtain money, or rather the social position, pleasures, and power, which money only can bestow. They are clear-headed, shrewd, practical men, well educated, because "it pays," and for the same reason they take good care to be supplied with the best instruments, and the latest literature. Many of them take up specialties because the work is easier, and the hours of labour are more under their control than in general practice. They strive to become connected with hospitals and medical schools, not for the love of mental exertion, or of science for its own sake, but as a respectable means of advertising, and of obtaining consultations. They write and lecture to keep their names before the public, and they must do both well, or fall behind in the race. They have the greater part of the valuable practice, and their writings, which constitute the greater part of our medical literature, are respectable in quality, and eminently useful.

They are the patrons of medical literature, the active working members of municipal medical societies, the men who are usually accepted as the representatives of the profession, not only here, but in all civilized countries; they may be famous physicians and great surgeons in the usual sense of the words, and as such, and only as such, should they receive the honour which is justly their due. They work for the present, and they have their reward in their own generation.

There is another large class, whose defects in general culture and in knowledge of the latest improvements in medicine, have been much dwelt upon by those disposed to take gloomy views of the condition of medical education in this country. The preliminary education of these physicians was defective, in some cases from lack of desire for it, but in the great majority from lack of opportunity, and their work in the medical school was confined to so much memorizing of text-books as was necessary to secure a diploma. In the course of practice they gradually obtain from personal experience, sometimes of a disagreeable kind, a knowledge of therapeutics, which enables them to treat the majority of their cases as successfully, perhaps, as their brethren more learned in theory. Occasionally they contribute a paper to a journal, or a report to a medical society; but they would rather talk than write, and find it very difficult to explain how or why they have succeeded, being like many excellent cooks in this respect. They are honest, conscientious, hard-working men, who are inclined to place great weight on their experience, and to be rather contemptuous of what they call "book learning and theories." To them our medical literature is indebted for a few interesting observations, and valuable suggestions in therapeutics, but for the most part, their experience, being unrecorded, has but a local usefulness.

These three classes have been referred to simply for the purpose of calling attention to the fact that, in speaking of "the physicians of the United States," it is necessary to be careful. There are many other classes, and they shade into each other and into empiricism in many ways. In discussions upon this subject, it seems to be often assumed that all physicians should possess the same qualifications, and be educated to the same standard, which, in one respect, is like saying that they should all be six feet high, and in another, is like the army regulations, which prescribe the same ration and allowance of clothing for Maine and Florida, Alaska and Oregon. A young and energetic man who has spent six years in obtaining a University education, and four more in the study of medicine as it ought to be studied, that is to say, in preparing himself to study and investigate for the rest of his life, will not settle in certain districts. He has invested ten years' labour, and from five to ten thousand dollars, and a locality which will give him a maximum income of, perhaps, fifteen hundred dollars per annum will not be satisfactory, in part because the capital should bring a better interest, in part because he will have acquired tastes which will make his life unpleasant in such places. Yet these places must have physicians of some sort, and it is not clear as to how they are to be supplied, if some of the universal and extensive reforms in medical education which have been proposed were to be enforced.

Certainly the standard for admission and for graduation at almost all our medical schools is too low, and one-half, at least, of these schools have no sufficient reason for existence; but it is not probable that it would improve matters much to establish a uniform, which must, of course, be a minimum, standard.

Of the material aids and instruments required for the advancement of medical science, such as hospitals, libraries, and museums, we have obtained as much as could be expected. With the proper use of those we now possess will come the demand for, and the supply of, still better facilities for the work of the scholar and observer.

The defects in American medicine are much the same as those observed in other branches of science in this country, and to a great extent are due to the same causes.

Culture, to flourish, requires appreciation and sympathy, to such an extent, at least, that its utterances shall not seem to its audience as if in an unknown tongue.

We have no reason to boast, or to be ashamed of what we have thus far accomplished; it has been but a little while since we have been furnished with the means of investigation needed to give our observations that accuracy and precision which alone can entitle medicine to a place among the sciences properly so called; and we may begin the new century in the hope and belief that to us applies the bright side of the maxim of Cousin, "It is better to have a future than a past."

CATALOGUE OF BOOKS

PUBLISHED BY

HENRY C. LEA.

(LATE LEA & BLANCHARD.)

The books in the annexed list will be sent by mail, post-paid, to any Post Office in the United States, on receipt of the printed prices. No risks of the mail, however, are assumed, either on money or books. Gentlemen will therefore, in most cases, find it more convenient to deal with the nearest bookseller.

Detailed catalogues furnished or sent free by mail on application. An illustrated catalogue of 64 octavo pages, handsomely printed, mailed on receipt of 10 cents. Address,

HENRY C. LEA,
Nos. 706 and 708 Sansom Street, Philadelphia.

PERIODICALS,

Free of Postage.

AMERICAN JOURNAL OF THE MEDICAL SCIENCES. Edited by Isaac Hays, M.D., published quarterly, about 1100 large 8vo. pages per annum,
MEDICAL NEWS AND LIBRARY, monthly, 384 large 8vo. pages per annum,
} For five Dollars per annum, in advance.

OR,

AMERICAN JOURNAL OF THE MEDICAL SCIENCES, Quarterly,
MEDICAL NEWS AND LIBRARY, monthly,
MONTHLY ABSTRACT OF MEDICAL SCIENCE, 48 pages per month, or nearly 600 pages per annum.
In all, about 2100 large 8vo. pages per annum,
} For six Dollars per annum, in advance.

MEDICAL NEWS AND LIBRARY, monthly, in advance, $1 00.

MONTHLY ABSTRACT OF MEDICAL SCIENCE, in advance, $2 50.

OBSTETRICAL JOURNAL. With an American Supplement, edited by J. V. INGHAM, M.D. $5 00 per annum, in advance. Single Numbers, 50 cents. Is published monthly, each number containing ninety-six octavo pages. Commencing with April, 1873.

ASHTON (T. J.) ON THE DISEASES, INJURIES, AND MALFORMATIONS OF THE RECTUM AND ANUS. With remarks on Habitual Constipation. Second American from the fourth London edition, with illustrations. 1 vol. 8vo. of about 300 pp. Cloth, $3 25.

ASHWELL (SAMUEL). A PRACTICAL TREATISE ON THE DISEASES OF WOMEN. Third American from the third London edition. In one 8vo. vol. of 528 pages. Cloth, $3 50.

ASHHURST (JOHN, Jr.) THE PRINCIPLES AND PRACTICE OF SURGERY. FOR THE USE OF STUDENTS AND PRACTITIONERS. In 1 large 8vo. vol. of over 1000 pages, containing 533 wood-cuts. Cloth, $6 50; leather, $7 50.

ATTFIELD (JOHN). CHEMISTRY; GENERAL, MEDICAL, AND PHARMACEUTICAL. Fifth edition, revised by the author. In 1 vol. 12mo. Cloth, $2 75; leather, $3 25.

ANDERSON (McCALL). ON THE TREATMENT OF DISEASES OF THE SKIN. In one small 8vo. vol. Cloth, $1 00.

BLOXAM (C. L.) CHEMISTRY, INORGANIC AND ORGANIC. With Experiments. In one handsome octavo volume of 700 pages, with 300 illustrations. Cloth, $4 00; leather, $5 00.

BRINTON (WILLIAM). LECTURES ON THE DISEASES OF THE STOMACH. From the second London edition, with illustrations. 1 vol. 8vo. Cloth, $3 25.

BRUNTON (T. LAUDER). A MANUAL OF MATERIA MEDICA AND THERAPEUTICS. In one 8vo. volume. (*Preparing.*)

BIGELOW (HENRY J.) ON DISLOCATION AND FRACTURE OF THE HIP, with the Reduction of the Dislocations by the Flexion Method. In one 8vo. vol. of 150 pp., with illustrations. Cloth, $2 50.

BASHAM (W. R.) RENAL DISEASES; A CLINICAL GUIDE TO THEIR DIAGNOSIS AND TREATMENT. With illustrations. 1 vol. 12mo. Cloth, $2 00.

BUMSTEAD (F. J.) THE PATHOLOGY AND TREATMENT OF VENEREAL DISEASES. Third edition, revised and enlarged, with illustrations. 1 vol. 8vo., of over 700 pages. Cloth, $5; leather, $6.

—— AND CULLERIER'S ATLAS OF VENEREAL. See "CULLERIER."

BARLOW (GEORGE H.) A MANUAL OF THE PRACTICE OF MEDICINE. 1 vol. 8vo., of over 600 pages. Cloth, $2 50.

BAIRD (ROBERT). IMPRESSIONS AND EXPERIENCES OF THE WEST INDIES. 1 vol. royal 12mo. Cloth, 75 cents.

BARNES (ROBERT). A PRACTICAL TREATISE ON THE DISEASES OF WOMEN. In one handsome 8vo. vol. of about 800 pages, with 169 illustrations. Cloth, $5; leather, $6. (*Lately issued.*)

BRYANT (THOMAS). THE PRACTICE OF SURGERY. In one handsome octavo volume, of over 1000 pages, with many illustrations. Cloth, $6 25; leather, $7 25. (*Lately issued.*)

BLANDFORD (G. FIELDING). INSANITY AND ITS TREATMENT. With an Appendix of the laws in force in the United States on the Confinement of the Insane, by Dr. Isaac Ray. In one handsome 8vo vol., of 471 pages. Cloth, $3 25.

BRISTOWE (JOHN SYER). A MANUAL OF THE PRACTICE OF MEDICINE. A new work, edited with additions by James H. Hutchinson, M.D. In one handsome 8vo. volume. (*In press.*)

BELLAMY'S MANUAL OF SURGICAL ANATOMY. With numerous illustrations. In one royal 12mo. vol. Cloth, $2 25. (*Lately issued.*)

BOWMAN (JOHN E.) A PRACTICAL HAND-BOOK OF MEDICAL CHEMISTRY. Sixth American, from the fourth London edition. With numerous illustrations. 1 vol. 12mo. of 350 pp. Cloth, $2 25.

—— INTRODUCTION TO PRACTICAL CHEMISTRY, INCLUDING ANALYSIS. Sixth American, from the sixth London edition, with numerous illustrations. 1 vol. 12mo. of 350 pp. Cloth, $2 25.

CARTER (R. BRUDENELL). A PRACTICAL TREATISE ON DISEASES OF THE EYE. With additions and test-types, by John Green, M.D. In one handsome 8vo. vol. of about 500 pages, with 124 illustrations. Cloth, $3 75. (*Just ready.*)

CHAMBERS (T. K.) A MANUAL OF DIET IN HEALTH AND DISEASE. In one handsome octavo volume of 310 pages. Cloth, $2 75. (*Just issued.*)

—— RESTORATIVE MEDICINE. An Harveian Annual Oration delivered at the Royal College of Physicians, London, June 21, 1871. In one small 12mo. volume. Cloth, $1 00.

COOPER (B. B.) LECTURES ON THE PRINCIPLES AND PRACTICE OF SURGERY. In one large 8vo. vol. of 750 pages. Cloth, $2 00.

CARPENTER (WM. B.) PRINCIPLES OF HUMAN PHYSIOLOGY, In one large vol. 8vo., of nearly 900 closely printed pages. Cloth, $5 50; leather, raised bands, $6 50.

—— PRIZE ESSAY ON THE USE OF ALCOHOLIC LIQUORS IN HEALTH AND DISEASE. New edition, with a Preface by D. F. Condie, M.D. 1 vol. 12mo. of 178 pages. Cloth, 60 cents.

CHRISTISON (ROBERT). DISPENSATORY OR COMMENTARY ON THE PHARMACOPŒIAS OF GREAT BRITAIN AND THE UNITED STATES. With a Supplement by R. E. Griffith. In one 8vo. vol. of over 1000 pages, containing 213 illustrations. Cloth, $4.

CHURCHILL (FLEETWOOD). ON THE THEORY AND PRACTICE OF MIDWIFERY. With notes and additions by D. Francis Condie, M.D. With about 200 illustrations. In one handsome 8vo. vol. of nearly 700 pages. Cloth, $4; leather, $5.

—— ESSAYS ON THE PUERPERAL FEVER, AND OTHER DISEASES PECULIAR TO WOMEN. In one neat octavo vol. of about 450 pages. Cloth, $2 50.

CONDIE (D. FRANCIS). A PRACTICAL TREATISE ON THE DISEASES OF CHILDREN. Sixth edition, revised and enlarged. In one large octavo volume of nearly 800 pages. Cloth, $5 25; leather, $6 25.

CULLERIER (A.) AN ATLAS OF VENEREAL DISEASES. Translated and edited by FREEMAN J. BUMSTEAD, M.D. A large imperial quarto volume, with 26 plates containing about 150 figures, beautifully colored, many of them the size of life. In one vol., strongly bound in cloth, $17.

—— Same work, in five parts, paper covers, for mailing, $3 per part.

CYCLOPEDIA OF PRACTICAL MEDICINE. By Dunglison, Forbes, Tweedie, and Conolly. In four large super-royal octavo volumes, of 3254 double-columned pages, leather, raised bands, $15. Cloth, $11.

CAMPBELL'S LIVES OF LORDS KENYON, ELLENBOROUGH, AND TENTERDEN. Being the third volume of "Campbell's Lives of the Chief Justices of England." In one crown octavo vol. Cloth, $2.

HENRY C. LEA'S PUBLICATIONS.

DALTON (J. C.) A TREATISE ON HUMAN PHYSIOLOGY. Sixth edition, thoroughly revised, and greatly enlarged and improved, with 316 illustrations. In one very handsome 8vo. vol. of 830 pp. Cloth, $5 50; leather, $6 50. (*Just issued.*)

DAVIS (F. H.) LECTURES ON CLINICAL MEDICINE. Second edition, revised and enlarged. In one 12mo. vol. Cloth, $1 75.

DON QUIXOTE DE LA MANCHA. Illustrated edition. In two handsome vols. crown 8vo. Cloth, $2 50; half morocco, $3 70.

DEWEES (W. P.) A TREATISE ON THE DISEASES OF FEMALES. With illustrations. In one 8vo. vol. of 536 pages. Cloth, $3.

—— A TREATISE ON THE PHYSICAL AND MEDICAL TREATMENT OF CHILDREN. In one 8vo. vol. of 548 pages. Cloth, $2 80.

DRUITT (ROBERT). THE PRINCIPLES AND PRACTICE OF MODERN SURGERY. A revised American, from the eighth London edition. Illustrated with 432 wood engravings. In one 8vo. vol. of nearly 700 pages. Cloth, $4; leather, $5.

DUNGLISON (ROBLEY). MEDICAL LEXICON; a Dictionary of Medical Science. Containing a concise explanation of the various subjects and terms of Anatomy, Physiology, Pathology, Hygiene, Therapeutics, Pharmacology, Pharmacy, Surgery, Obstetrics, Medical Jurisprudence, and Dentistry. Notices of Climate and of Mineral Waters; Formulæ for Officinal, Empirical, and Dietetic Preparations, with the accentuation and Etymology of the Terms, and the French and other Synonymes. In one very large royal 8vo. vol. New edition. Cloth, $6 50; leather, $7 50. (*Just issued.*)

—— HUMAN PHYSIOLOGY. Eighth edition, thoroughly revised. In two large 8vo. vols. of about 1500 pp., with 532 illus. Cloth, $7.

—— NEW REMEDIES, WITH FORMULÆ FOR THEIR PREPARATION AND ADMINISTRATION. Seventh edition. In one very large 8vo. vol. of 770 pages. Cloth, $4.

DE LA BECHE'S GEOLOGICAL OBSERVER. In one large 8vo. vol. of 700 pages, with 300 illustrations. Cloth, $4.

DANA (JAMES D.) THE STRUCTURE AND CLASSIFICATION OF ZOOPHYTES. With illustrations on wood. In one imperial 4to. vol. Cloth, $4 00.

ELLIS (BENJAMIN). THE MEDICAL FORMULARY. Being a collection of prescriptions derived from the writings and practice of the most eminent physicians of America and Europe. Twelfth edition, carefully revised by A. H. Smith, M. D. In one 8vo. volume of 374 pages. Cloth, $3.

ERICHSEN (JOHN). THE SCIENCE AND ART OF SURGERY. A new and improved American, from the sixth enlarged and revised London edition. Illustrated with 630 engravings on wood. In two large 8vo. vols. Cloth, $9 00; leather, raised bands, $11 00.

ENCYCLOPÆDIA OF GEOGRAPHY. In three large 8vo. vols. Illustrated with 83 maps and about 1100 wood-cuts. Cloth, $5.

FOTHERGILL'S PRACTITIONER'S HANDBOOK OF TREATMENT. In one handsome octavo volume. (*In preparation for early publication.*)

FENWICK (SAMUEL). THE STUDENTS' GUIDE TO MEDICAL DIAGNOSIS. From the Third Revised and Enlarged London Edition. In one vol. royal 12mo., with numerous illustrations. Cloth, $2 25. (*Just issued.*)

FLETCHER'S NOTES FROM NINEVEH, AND TRAVELS IN MESOPOTAMIA, ASSYRIA, AND SYRIA. In one 12mo. vol. Cloth, 75 cts.

FOX ON DISEASES OF THE STOMACH. From the third London edition. In one octavo vol. Cloth, $2. (*Just issued.*)

FOX (TILBURY). EPITOME OF SKIN DISEASES, with Formulæ for Students and Practitioners. In one small 12mo. vol. (*In press.*)

FLINT (AUSTIN). A TREATISE ON THE PRINCIPLES AND PRACTICE OF MEDICINE. Fourth edition, thoroughly revised and enlarged. In one large 8vo. volume of 1070 pages. Cloth, $6; leather, raised bands, $7. (*Just issued.*)

——— A PRACTICAL TREATISE ON THE PHYSICAL EXPLORATION OF THE CHEST, AND THE DIAGNOSIS OF DISEASES AFFECTING THE RESPIRATORY ORGANS. Second and revised edition. One 8vo. vol. of 595 pages. Cloth, $4 50.

——— A PRACTICAL TREATISE ON THE DIAGNOSIS AND TREATMENT OF DISEASES OF THE HEART. Second edition, enlarged. In one neat 8vo. vol. of over 500 pages, $4 00.

——— ON PHTHISIS: ITS MORBID ANATOMY, ETIOLOGY, ETC., in a series of Clinical Lectures. A new work. In one handsome 8vo. volume. Cloth, $3 50. (*Just issued.*)

——— A MANUAL OF PERCUSSION AND AUSCULTATION; of the Physical Diagnosis of Diseases of the Lungs and Heart, and of Thoracic Aneurism. In one handsome royal 12mo. volume. Cloth, $1 75. (*Now ready.*)

——— MEDICAL ESSAYS. In one neat 12mo. volume. Cloth, $1 38.

FOWNES (GEORGE). A MANUAL OF ELEMENTARY CHEMISTRY. From the tenth enlarged English edition. In one royal 12mo. vol. of 857 pages, with 197 illustrations. Cloth, $2 75; leather, $3 25.

FULLER (HENRY). ON DISEASES OF THE LUNGS AND AIR PASSAGES. Their Pathology, Physical Diagnosis, Symptoms, and Treatment. From the second English edition. In one 8vo. vol. of about 500 pages. Cloth, $3 50.

GALLOWAY (ROBERT). A MANUAL OF QUALITATIVE ANALYSIS. From the fifth English edition. In one 12mo. vol. Cloth, $2 50. (*Lately published.*)

GLUGE (GOTTLIEB). ATLAS OF PATHOLOGICAL HISTOLOGY. Translated by Joseph Leidy, M.D., Professor of Anatomy in the University of Pennsylvania, &c. In one vol. imperial quarto, with 320 copperplate figures, plain and colored. Cloth, $4.

GREEN (T. HENRY). AN INTRODUCTION TO PATHOLOGY AND MORBID ANATOMY. Second Amer., from the third Lond. Ed. In one handsome 8vo. vol., with numerous illustrations. Cloth, $2 75. (*Just issued*)

GRAY (HENRY). ANATOMY, DESCRIPTIVE AND SURGICAL. A new American, from the fifth and enlarged London edition. In one large imperial 8vo. vol. of about 900 pages, with 462 large and elaborate engravings on wood. Cloth, $6; leather, $7. (*Lately issued.*)

GRIFFITH (ROBERT E.). A UNIVERSAL FORMULARY, CONTAINING THE METHODS OF PREPARING AND ADMINISTERING OFFICINAL AND OTHER MEDICINES. Third and Enlarged edition. Edited by John M. Maisch. In one large 8vo. vol. of 800 pages, double columns. Cloth, $4 50; leather, $5 50.

GROSS (SAMUEL D.) A SYSTEM OF SURGERY, PATHOLOGICAL, DIAGNOSTIC, THERAPEUTIC, AND OPERATIVE. Illustrated by 1403 engravings. Fifth edition, revised and improved. In two large imperial 8vo. vols. of over 2200 pages, strongly bound in leather, raised bands, $15.

GROSS (SAMUEL D.) A PRACTICAL TREATISE ON THE DISeases, Injuries, and Malformations of the Urinary Bladder, the Prostate Gland, and the Urethra. Third Edition, thoroughly Revised and Condensed, by Samuel W. Gross, M.D., Surgeon to the Philadelphia Hospital. In one handsome octavo volume, with about two hundred illustrations. Cloth, $4 50. (*Now ready.*)

—— A PRACTICAL TREATISE ON FOREIGN BODIES IN THE AIR PASSAGES. In one 8vo. vol. of 468 pages. Cloth, $2 75.

GIBSON'S INSTITUTES AND PRACTICE OF SURGERY. In two 8vo. vols. of about 1000 pages, leather, $6 50.

GOSSELIN (L) CLINICAL LECTURES ON SURGERY, Delivered at the Hospital of La Charité Translated from the French by Lewis A. Stimson, M.D., Surgeon to the Presbyterian Hospital, New York. With illustrations. (*Publishing in the Medical News and Library for* 1876-7.)

HUDSON (A.) LECTURES ON THE STUDY OF FEVER. 1 vol. 8vo., 316 pages. Cloth, $2 50.

HEATH (CHRISTOPHER). PRACTICAL ANATOMY; A MANUAL OF DISSECTIONS. With additions, by W. W. Keen, M. D. In 1 volume; with 247 illustrations. Cloth, $3 50; leather, $4.

HARTSHORNE (HENRY). ESSENTIALS OF THE PRINCIPLES AND PRACTICE OF MEDICINE. Fourth and revised edition. In one 12mo. vol. Cloth, $2 63; half bound, $2 88. (*Lately issued*)

—— CONSPECTUS OF THE MEDICAL SCIENCES. Comprising Manuals of Anatomy, Physiology, Chemistry, Materia Medica, Practice of Medicine, Surgery, and Obstetrics. Second Edition. In one royal 12mo. volume of over 1000 pages, with 477 illustrations. Strongly bound in leather, $5 00; cloth, $4 25. (*Lately issued.*)

—— A HANDBOOK OF ANATOMY AND PHYSIOLOGY. In one neat royal 12mo. volume, with many illustrations. Cloth, $1 75.

HAMILTON (FRANK H.) A PRACTICAL TREATISE ON FRACTURES AND DISLOCATIONS. Fifth edition, carefully revised. In one handsome 8vo. vol. of 830 pages, with 344 illustrations. Cloth, $5 75; leather, $6 75. (*Just issued.*)

HOLMES (TIMOTHY). SURGERY, ITS PRINCIPLES AND PRACTICE. In one handsome 8vo. volume of 1000 pages, with 411 illustrations. Cloth, $6; leather, with raised bands, $7. (*Just ready.*)

HOBLYN (RICHARD D.) A DICTIONARY OF THE TERMS USED IN MEDICINE AND THE COLLATERAL SCIENCES In one 12mo. volume, of over 500 double-columned pages. Cloth, $1 50; leather, $2.

HODGE (HUGH L.) ON DISEASES PECULIAR TO WOMEN, INCLUDING DISPLACEMENTS OF THE UTERUS. Second and revised edition. In one 8vo. volume. Cloth, $4 50.

—— THE PRINCIPLES AND PRACTICE OF OBSTETRICS. Illustrated with large lithographic plates containing 159 figures from original photographs, and with numerous wood-cuts. In one large quarto vol. of 550 double-columned pages. Strongly bound in cloth, $14.

HOLLAND (SIR HENRY). MEDICAL NOTES AND REFLECTIONS. From the third English edition. In one 8vo. vol. of about 500 pages. Cloth, $3 50.

HODGES (RICHARD M.) PRACTICAL DISSECTIONS. Second edition. In one neat royal 12mo. vol., half bound, $2.

HUGHES. SCRIPTURE GEOGRAPHY AND HISTORY, with 12 colored maps. In 1 vol. 12mo. Cloth, $1.

HORNER (WILLIAM E.) SPECIAL ANATOMY AND HISTOLOGY. Eighth edition, revised and modified. In two large 8vo. vols. of over 1000 pages, containing 300 wood-cuts. Cloth, $6.

HILL (BERKELEY). SYPHILIS AND LOCAL CONTAGIOUS DISORDERS. In one 8vo. volume of 467 pages. Cloth, $3 25.

HILLIER (THOMAS). HAND-BOOK OF SKIN DISEASES. Second Edition. In one neat royal 12mo. volume of about 300 pp., with two plates. Cloth, $2 25.

HALL (MRS. M.) LIVES OF THE QUEENS OF ENGLAND BEFORE THE NORMAN CONQUEST. In one handsome 8vo. vol. Cloth, $2 25; crimson cloth, $2 50; half morocco, $3.

JONES (C. HANDFIELD). CLINICAL OBSERVATIONS ON FUNCTIONAL NERVOUS DISORDERS. Second American Edition. In one 8vo. vol. of 348 pages. Cloth, $3 25.

KIRKES (WILLIAM SENHOUSE). A MANUAL OF PHYSIOLOGY. A new American, from the eighth London edition. One vol., with many illus., 12mo. Cloth, $3 25; leather, $3 75.

KNAPP (F.) TECHNOLOGY; OR CHEMISTRY, APPLIED TO THE ARTS AND TO MANUFACTURES, with American additions, by Prof. Walter R. Johnson. In two 8vo. vols., with 500 ill. Cloth, $6.

KENNEDY'S MEMOIRS OF THE LIFE OF WILLIAM WIRT. In two vols. 12mo. Cloth, $2.

LEA (HENRY C.) SUPERSTITION AND FORCE; ESSAYS ON THE WAGER OF LAW, THE WAGER OF BATTLE, THE ORDEAL, AND TORTURE. Second edition, revised. In one handsome royal 12mo. vol., $2 75.

—— STUDIES IN CHURCH HISTORY. The Rise of the Temporal Power—Benefit of Clergy—Excommunication. In one handsome 12mo. vol. of 515 pp. Cloth, $2 75.

—— AN HISTORICAL SKETCH OF SACERDOTAL CELIBACY IN THE CHRISTIAN CHURCH. In one handsome octavo volume of 602 pages. Cloth, $3 75.

LA ROCHE (R.) YELLOW FEVER. In two 8vo. vols. of nearly 1500 pages. Cloth, $7.

—— PNEUMONIA. In one 8vo. vol. of 500 pages. Cloth, $3.

LINCOLN (D. F.) ELECTRO-THERAPEUTICS. A Condensed Manual of Medical Electricity. In one neat royal 12mo. volume, with illustrations. Cloth, $1 50. (*Just issued.*)

LEISHMAN (WILLIAM). A SYSTEM OF MIDWIFERY. Including the Diseases of Pregnancy and the Puerperal State. Second American, from the Second English Edition. With additions, by J. S. Parry, M.D. In one very handsome 8vo. vol. of 800 pages and 200 illustrations. Cloth, $5; leather, $6. (*Just issued.*)

LAURENCE (J. Z.) AND MOON (ROBERT C.) A HANDY-BOOK OF OPHTHALMIC SURGERY. Second edition, revised by Mr. Laurence. With numerous illus. In one 8vo. vol. Cloth, $2 75.

LEHMANN (C. G.) PHYSIOLOGICAL CHEMISTRY. Translated by George F. Day, M. D. With plates, and nearly 200 illustrations. In two large 8vo. vols., containing 1200 pages. Cloth, $6.

—— A MANUAL OF CHEMICAL PHYSIOLOGY. In one very handsome 8vo. vol. of 336 pages. Cloth, $2 25.

LAWSON (GEORGE). INJURIES OF THE EYE, ORBIT, AND EYE-LIDS, with about 100 illustrations. From the last English edition. In one handsome 8vo. vol. Cloth, $3 50.

LUDLOW (J. L.) A MANUAL OF EXAMINATIONS UPON ANATOMY, PHYSIOLOGY, SURGERY, PRACTICE OF MEDICINE, OBSTETRICS, MATERIA MEDICA, CHEMISTRY, PHARMACY, AND THERAPEUTICS. To which is added a Medical Formulary. Third edition. In one royal 12mo. vol. of over 800 pages. Cloth, $3 25; leather, $3 75.

LAYCOCK (THOMAS). LECTURES ON THE PRINCIPLES AND METHODS OF MEDICAL OBSERVATION AND RESEARCH. In one 12mo. vol. Cloth, $1.

LYNCH (W. F.) A NARRATIVE OF THE UNITED STATES EXPEDITION TO THE DEAD SEA AND RIVER JORDAN. In one large octavo vol., with 28 beautiful plates and two maps. Cloth, $3.

—— Same Work, condensed edition. One vol. royal 12mo. Cloth, $1.

LEE (HENRY) ON SYPHILIS. In one 8vo. vol. Cloth, $2 25.

LYONS (ROBERT D.) A TREATISE ON FEVER. In one neat 8vo. vol. of 362 pages. Cloth, $2 25.

MARSHALL (JOHN). OUTLINES OF PHYSIOLOGY, HUMAN AND COMPARATIVE. With Additions by FRANCIS G. SMITH. M. D., Professor of the Institutes of Medicine in the University of Pennsylvania. In one 8vo. volume of 1026 pages, with 122 illustrations. Strongly bound in leather, raised bands, $7 50. Cloth, $6 50.

MACLISE (JOSEPH). SURGICAL ANATOMY. In one large imperial quarto vol., with 68 splendid plates, beautifully colored; containing 190 figures, many of them life size. Cloth, $14.

MEIGS (CHAS. D.). ON THE NATURE, SIGNS, AND TREATMENT OF CHILDBED FEVER. In one 8vo. vol. of 365 pages. Cloth, $2.

MILLER (JAMES). PRINCIPLES OF SURGERY. Fourth American, from the third Edinburgh edition. In one large 8vo. vol. of 700 pages, with 240 illustrations. Cloth, $3 75.

—— THE PRACTICE OF SURGERY. Fourth American, from the last Edinburgh edition. In one large 8vo. vol. of 700 pages, with 364 illustrations. Cloth, $3 75.

MONTGOMERY (W. F.) AN EXPOSITION OF THE SIGNS AND SYMPTOMS OF PREGNANCY. From the second English edition. In one handsome 8vo. vol. of nearly 600 pages. Cloth, $3 75.

MULLER (J.) PRINCIPLES OF PHYSICS AND METEOROLOGY. In one large 8vo. vol. with 550 wood-cuts, and two colored plates. Cloth, $4 50.

MIRABEAU; A LIFE HISTORY. In one 12mo. vol. Cloth, 75 cts.

MACFARLAND'S TURKEY AND ITS DESTINY. In 2 vols. royal 12mo. Cloth, $2.

MARSH (MRS.) A HISTORY OF THE PROTESTANT REFORMATION IN FRANCE. In 2 vols. royal 12mo. Cloth, $2.

NELIGAN (J. MOORE). AN ATLAS OF CUTANEOUS DISEASES. In one quarto volume, with beautifully colored plates, &c. Cloth, $5 50.

NEILL (JOHN) AND SMITH (FRANCIS G.) COMPENDIUM OF THE VARIOUS BRANCHES OF MEDICAL SCIENCE. In one handsome 12mo. vol. of about 1000 pages, with 374 wood-cuts. Cloth, $4; leather, raised bands, $4 75.

NIEBUHR (B. G.) LECTURES ON ANCIENT HISTORY; comprising the history of the Asiatic Nations, the Egyptians, Greeks, Macedonians, and Carthagenians. Translated by Dr. L. Schmitz. In three neat volumes, crown octavo. Cloth, $5 00.

ODLING (WILLIAM). A COURSE OF PRACTICAL CHEMISTRY FOR THE USE OF MEDICAL STUDENTS. From the fourth revised London edition. In one 12mo. vol. of 261 pp., with 75 illustrations. Cloth, $2.

PLAYFAIR (W. S) A TREATISE ON THE SCIENCE AND PRACTICE OF MIDWIFERY. In one handsome octavo vol. of 576 pp., with 166 illustrations. Cloth, $4; leather, $5. (*Now ready.*)

PAVY (F. W.) A TREATISE ON THE FUNCTION OF DIGESTION, ITS DISORDERS AND THEIR TREATMENT. From the second London ed. In one 8vo. vol. of 246 pp. Cloth, $2. (*Lately issued.*)

―――― A TREATISE ON FOOD AND DIETETICS, PHYSIOLOGICALLY AND THERAPEUTICALLY CONSIDERED. In one neat octavo volume of about 500 pages. Cloth, $4 75. (*Just issued.*)

PARRISH (EDWARD). A TREATISE ON PHARMACY. With many Formulæ and Prescriptions. Fourth edition. Enlarged and thoroughly revised by Thomas S. Wiegand. In one handsome 8vo. vol. of 977 pages, with 280 illus. Cloth, $5 50; leather, $6 50. (*Just issued.*)

PIRRIE (WILLIAM) THE PRINCIPLES AND PRACTICE OF SURGERY. In one handsome octavo volume of 780 pages, with 316 illustrations. Cloth, $3 75.

PEREIRA (JONATHAN). MATERIA MEDICA AND THERAPEUTICS. An abridged edition. With numerous additions and references to the United States Pharmacopœia. By Horatio C. Wood, M. D. In one large octavo volume, of 1040 pages, with 236 illustrations. Cloth, $7 00; leather, raised bands, $8 00.

PULSZKY'S MEMOIRS OF AN HUNGARIAN LADY. In one neat royal 12mo. vol. Cloth, $1.

PAGET'S HUNGARY AND TRANSYLVANIA. In two royal 12mo. vols. Cloth, $2.

ROBERTS (WILLIAM). A PRACTICAL TREATISE ON URINARY AND RENAL DISEASES. A second American, from the second London edition. With numerous illustrations and a colored plate. In one very handsome 8vo. vol. of 616 pages. Cloth, $4 50. (*Just issued.*)

RAMSBOTHAM (FRANCIS H.) THE PRINCIPLES AND PRACTICE OF OBSTETRIC MEDICINE AND SURGERY. In one imperial 8vo. vol. of 650 pages, with 64 plates, besides numerous woodcuts in the text. Strongly bound in leather, $7.

RIGBY (EDWARD). A SYSTEM OF MIDWIFERY. Second American edition. In one handsome 8vo. vol. of 422 pages. Cloth, $2 50.

RANKE'S HISTORY OF THE TURKISH AND SPANISH EMPIRES in the 16th and beginning of 17th Century. In one 8vo. volume, paper, 25 cts.

―――― HISTORY OF THE REFORMATION IN GERMANY. Parts I., II., III. In one vol. Cloth, $1.

SMITH (EUSTACE). ON THE WASTING DISEASES OF CHILDREN. Second American edition, enlarged. In one 8vo. vol. Cloth, $2 50. (*Just issued.*)

SARGENT (F. W.) ON BANDAGING AND OTHER OPERATIONS OF MINOR SURGERY. New edition, with an additional chapter on Military Surgery. In one handsome royal 12mo. vol. of nearly 400 pages, with 184 wood-cuts. Cloth, $1 75.

SMITH (J. LEWIS.) A TREATISE ON THE DISEASES OF INFANCY AND CHILDHOOD. Third Edition, revised and enlarged. In one large 8vo. volume of 724 pages, with illustrations. Cloth, $5; leather, $6. (*Just issued.*)

SHARPEY (WILLIAM) AND QUAIN (JONES AND RICHARD). HUMAN ANATOMY. With notes and additions by Jos. Leidy, M. D., Prof. of Anatomy in the University of Pennsylvania. In two large 8vo. vols. of about 1300 pages, with 511 illustrations. Cloth, $6.

SKEY (FREDERIC C.) OPERATIVE SURGERY. In one 8vo. vol. of over 650 pages, with about 100 wood-cuts. Cloth, $3 25.

SLADE (D. D.) DIPHTHERIA; ITS NATURE AND TREATMENT. Second edition. In one neat royal 12mo. vol. Cloth, $1 25.

SMITH (HENRY H.) AND HORNER (WILLIAM E.) ANATOMICAL ATLAS. Illustrative of the structure of the Human Body. In one large imperial 8vo. vol., with about 650 beautiful figures. Cloth, $4 50.

SMITH (EDWARD). CONSUMPTION; ITS EARLY AND REMEDIABLE STAGES. In one 8vo. vol. of 254 pp. Cloth, $2 25.

STILLE (ALFRED). THERAPEUTICS AND MATERIA MEDICA. Fourth edition, revised and enlarged. In two large and handsome volumes 8vo. Cloth, $10; leather, $12. (*Just issued.*)

SCHMITZ AND ZUMPT'S CLASSICAL SERIES. In royal 18mo. CORNELII NEPOTIS LIBER DE EXCELLENTIBUS DUCIBUS EXTERARUM GENTIUM, CUM VITIS CATONIS ET ATTICI. With notes, &c. Price in cloth, 60 cents; half bound, 70 cts.

C. I. CÆSARIS COMMENTARII DE BELLO GALLICO. With notes, map, and other illustrations. Price in cloth, 60 cents; half bound, 70 cents.

C. C. SALLUSTII DE BELLO CATILINARIO ET JUGURTHINO. With notes, map, &c. Price in cloth, 60 cents; half bound, 70 cents.

Q. CURTII RUFII DE GESTIS ALEXANDRI MAGNI LIBRI VIII. With notes, map, &c. Price in cloth, 80 cents; half bound, 90 cents.

P. VIRGILII MARONIS CARMINA OMNIA. Price in cloth, 85 cents; half bound, $1.

M. T. CICERONIS ORATIONES SELECTÆ XII. With notes, &c. Price in cloth, 70 cents; half bound, 80 cents.

ECLOGÆ EX Q. HORATII FLACCI POEMATIBUS. With notes, &c. Price in cloth, 70 cents; half bound, 80 cents.

ADVANCED LATIN EXERCISES, WITH SELECTIONS FOR READING. Revised, with additions. Cloth, price 60 cents; half bound, 70 cents.

SWAYNE (JOSEPH GRIFFITHS). OBSTETRIC APHORISMS. A new American, from the fifth revised English edition. With additions by E. R. Hutchins, M. D. In one small 12mo. vol. of 177 pp., with illustrations. Cloth, $1 25.

STURGES (OCTAVIUS). AN INTRODUCTION TO THE STUDY OF CLINICAL MEDICINE. In one 12mo. vol. Cloth, $1 25.

SCHOEDLER (FREDERICK) AND MEDLOCK (HENRY). WONDERS OF NATURE. An elementary introduction to the Sciences of Physics, Astronomy, Chemistry, Mineralogy, Geology, Botany, Zoology, and Physiology. Translated from the German by H. Medlock. In one neat 8vo. vol., with 679 illustrations. Cloth, $3.

STOKES (W.) LECTURES ON FEVER. In one 8vo. vol. Cloth, $2.

SMALL BOOKS ON GREAT SUBJECTS. Twelve works; each one 10 cents, sewed, forming a neat and cheap series; or done up in 3 vols., cloth, $1 50.

STRICKLAND (AGNES). LIVES OF THE QUEENS OF HENRY THE VIII. AND OF HIS MOTHER. In one crown octavo vol., extra cloth, $1; black cloth, 90 cents.

——— MEMOIRS OF ELIZABETH, SECOND QUEEN REGNANT OF ENGLAND AND IRELAND. In one crown octavo vol., extra cloth, $1 40; black cloth, $1 30.

TANNER (THOMAS HAWKES). A MANUAL OF CLINICAL MEDICINE AND PHYSICAL DIAGNOSIS. Third American from the second revised English edition. Edited by Tilbury Fox, M. D. In one handsome 12mo. volume of 366 pp. Cloth, $1 50.

——— ON THE SIGNS AND DISEASES OF PREGNANCY. From the second English edition. With four colored plates and numerous illustrations on wood. In one vol. 8vo. of about 500 pages. Cloth, $4 25.

TUKE (DANIEL HACK). INFLUENCE OF THE MIND UPON THE BODY. In one handsome 8vo. vol. of 416 pp. Cloth, $3 25. (*Just issued.*)

TAYLOR (ALFRED S.) MEDICAL JURISPRUDENCE. Seventh American edition. Edited by John J. Reese, M.D. In one large 8vo. volume of 879 pages. Cloth, $5; leather, $6. (*Just issued.*)

——— PRINCIPLES AND PRACTICE OF MEDICAL JURISPRUDENCE. From the Second English Edition. In two large 8vo. vols. Cloth, $10; leather, $12. (*Just issued.*)

——— ON POISONS IN RELATION TO MEDICINE AND MEDICAL JURISPRUDENCE. Third American from the Third London Edition. 1 vol. 8vo. of 788 pages, with 104 illustrations. Cloth, $5 50; leather, $6 50. (*Just issued.*)

THOMAS (T. GAILLARD). A PRACTICAL TREATISE ON THE DISEASES OF FEMALES. Fourth edition, thoroughly revised. In one large and handsome octavo volume of 801 pages, with 191 illustrations. Cloth, $5 00; leather, $6 00. (*Just issued.*)

TODD (ROBERT BENTLEY). CLINICAL LECTURES ON CERTAIN ACUTE DISEASES. In one vol. 8vo. of 320 pp., cloth, $2 50.

THOMPSON (SIR HENRY). CLINICAL LECTURES ON DISEASES OF THE URINARY ORGANS. Second and revised edition. In one 8vo. volume, with illustrations. Cloth, $2 25. (*Just issued.*)

——— THE PATHOLOGY AND TREATMENT OF STRICTURE OF THE URETHRA AND URINARY FISTULÆ. From the third English edition. In one 8vo. vol. of 359 pp., with illus. Cloth, $3 50.

——— THE DISEASES OF THE PROSTATE, THEIR PATHOLOGY AND TREATMENT. Fourth edition, revised. In one very handsome 8vo. vol. of 355 pp., with 13 plates. Cloth, $3 75.

WALSHE (W. H.) PRACTICAL TREATISE ON THE DISEASES OF THE HEART AND GREAT VESSELS. Third American from the third revised London edition. In one 8vo. vol. of 420 pages. Cloth, $3.

WATSON (THOMAS). LECTURES ON THE PRINCIPLES AND PRACTICE OF PHYSIC. A new American from the fifth and enlarged English edition, with additions by H. Hartshorne, M.D. In two large and handsome octavo volumes. Cloth, $9; leather, $11.

WÖHLER'S OUTLINES OF ORGANIC CHEMISTRY. Translated from the 8th German edition, by Ira Remsen, M.D. In one neat 12mo. vol. Cloth, $3 00. (*Lately issued.*)

WELLS (J. SOELBERG). A TREATISE ON THE DISEASES OF THE EYE. Second American, from the Third English edition, with additions by I. Minis Hays, M.D. In one large and handsome octavo vol., with 6 colored plates and many wood-cuts, also selections from the test-types of Jaeger and Snellen. Cloth, $5 00; leather, $6 00.

WHAT TO OBSERVE AT THE BEDSIDE AND AFTER DEATH IN MEDICAL CASES. In one royal 12mo. vol. Cloth, $1.

WEST (CHARLES). LECTURES ON THE DISEASES PECULIAR TO WOMEN. Third American from the Third English edition. In one octavo volume of 550 pages. Cloth, $3 75; leather, $4 75.

—— LECTURES ON THE DISEASES OF INFANCY AND CHILDHOOD. Fifth American from the sixth revised English edition. In one large 8vo. vol. of 670 closely printed pages. Cloth, $4 50; leather, $5 50. (*Just issued.*)

—— ON SOME DISORDERS OF THE NERVOUS SYSTEM IN CHILDHOOD. From the London Edition. In one small 12mo. volume. Cloth, $1.

—— AN ENQUIRY INTO THE PATHOLOGICAL IMPORTANCE OF ULCERATION OF THE OS UTERI. In one vol. 8vo. Cloth, $1 25.

WILLIAMS (CHARLES J. B. and C. T.) PULMONARY CONSUMPTION: ITS NATURE, VARIETIES, AND TREATMENT. In one neat octavo volume. Cloth, $2 50. (*Lately published.*)

WILSON (ERASMUS). A SYSTEM OF HUMAN ANATOMY. A new and revised American from the last English edition. Illustrated with 397 engravings on wood. In one handsome 8vo. vol. of over 600 pages. Cloth, $4; leather, $5.

—— ON DISEASES OF THE SKIN. The seventh American from the last English edition. In one large 8vo. vol. of over 800 pages. Cloth, $5.

Also, A SERIES OF PLATES, illustrating "Wilson on Diseases of the Skin," consisting of 20 plates, thirteen of which are beautifully colored, representing about one hundred varieties of Disease. $5 50.

Also, the TEXT AND PLATES, bound in one volume. Cloth, $10.

—— THE STUDENT'S BOOK OF CUTANEOUS MEDICINE. In one handsome royal 12mo. vol. Cloth, $3 50.

WINSLOW (FORBES). ON OBSCURE DISEASES OF THE BRAIN AND DISORDERS OF THE MIND. In one handsome 8vo. vol. of nearly 600 pages. Cloth, $4 25.

WINCKEL ON PATHOLOGY AND TREATMENT OF CHILDBED. With Additions by the Author. Translated by Chadwick. In one handsome octavo volume of 484 pages. Cloth, $4. (*Just issued.*)

ZEISSL ON VENEREAL DISEASES. Translated by Sturgis. (*Preparing.*)

www.ingramcontent.com/pod-product-compliance
Lightning Source LLC
Chambersburg PA
CBHW032030220426
43664CB00006B/419